T0135575

Control of Medium-Voltage Drives
at Very Low Switching Frequency

vom Fachbereich
Elektrotechnik, Informationstechnik und Medientechnik
der Bergischen Universität Wuppertal
zur Erlangung des akademischen Grades eines
Doktor-Ingenieurs
genehmigte Dissertation

vorgelegt von
Diplom-Ingenieur Nikolaos Oikonomou
aus Thessaloniki, Griechenland

Referent: Prof. Dr.-Ing. Joachim Holtz
Korreferent: Prof. Dr.-Ing. Mario Pacas

Tag der mündlichen Prüfung: 08. Februar 2008

Bibliografische Information der Deutschen Nationalbibliothek

Die Deutsche Nationalbibliothek verzeichnet diese Publikation in der
Deutschen Nationalbibliografie; detaillierte bibliografische Daten sind
im Internet über http://dnb.d-nb.de abrufbar.

ISBN 978-3-8325-1990-2

Logos Verlag Berlin GmbH
Comeniushof, Gubener Str. 47,
10243 Berlin
Tel.: +49 030 42 85 10 90
Fax: +49 030 42 85 10 92
INTERNET: http://www.logos-verlag.de

Contents

Acknowledgments

In producing the present work I was fortunate to receive the guidance and assistance of Prof. Dr.-Ing. Joachim Holtz. He was the one that supervised this work throughout all its stages; its completion would have not been possible without his leadership. He provided the motivation to convert new lines of thought into practical schemes and results and he bid his accumulated scientific experience in order to resolve intriguing problems that were encountered during the procedure. I had the opportunity to profit from the daily interaction in the laboratory with him; his thorough way to approach issues related to academic and industrial research acted as a valuable example to me.

Prof. Dr.-Ing. Ralph Kennel was the Head of the Electrical Machines and Drives Laboratories of the Wuppertal University during the time I worked there. He provided me with access to the laboratory infrastructure and with the opportunity to work as a scientific collaborator parallel to my research activities. Furthermore, he contributed with constructive advice on issues related to the present work and he retrieved financial means to allow me attend numerous Conferences and workshops. I thank him for all his valuable support and for his caring attitude.

I also wish to express my gratitude towards Prof. Dr.-Ing. Mario Pacas of the University of Siegen for taking up the co-examination. The corrections he proposed helped to improve the quality level of the present manuscript.

The work that was conducted in the laboratory in Wuppertal was also tested in an industrial environment, this of WEG Automaçao Ltd. in Brazil: the support of Dipl.-Ing. Norton Petry, Dipl.-Ing. Paulo Jose Torri, and Dipl.-Ing. Gilberto da Cunha was decisive. They shared their wide knowledge on industry applications with me and they put intensive effort in the evaluation of the performance of the control schemes that were developed. I wish to thank them for their valuable contribution to the work and for their considerate personal approach that I had the chance to benefit from.

The staff members – both scientific and technical – form the kernel of the Electrical Machines and Drives Laboratories of the University of Wuppertal; the quality of their work defines the overall level of the institute. I feel indebted to all my former and present colleagues, because I profited from them in both a professional and in an interpersonal way. Collaborating with them in the laboratory provided me with many useful ideas and the ability to acquire an overview of various aspect of our research field. At the same time, I enjoyed the advantage of developing friendships that extend beyond the timetable.

At last, but not least, I wish to mention those persons that stand closest to me: my parents Maria and Konstantinos Oikonomou, my brother Evangelos, and my partner in life, Theodora Arabatzi. There exist not enough words to describe my gratitude for their love and patience.

1. Introduction

Medium-voltage ac drives based on voltage source inverters are in increasing demand for numerous industrial applications. Higher power ratings of the inverter are achieved by increasing the voltage rather than the current, which provides better efficiency. The three-level inverter is then a preferred choice for medium-voltage drive applications: it allows operation at double the dc-link voltage and lower harmonic distortion of the ac currents as compared to the conventional two-level inverter [1].

Of particular interest is the three-level, neutral-point clamped (NPC) topology [2], which has been recently established as an industrial standard in the medium-voltage range. Among issues related to control methods of the NPC inverter, the neutral point potential offset [3] has received increasing attention. Existing neutral point potential balancing methods are reviewed, and an improved strategy is subsequently defined and analyzed.

The requirement of high dynamic performance can be met by employing vector control schemes; these provide large control bandwidth and decoupled control of torque and excitation.

It is desirable to operate medium-voltage inverters at very low switching frequency. This reduces the switching losses of the power semiconductors that contribute a major portion of the total device losses. It is a disadvantage, though, that the signal delays within the control loop increase as the switching frequency reduces [4]. The control bandwidth then decreases, and dynamic decoupling between torque and flux becomes less effective. The undesired cross-coupling effect cannot be adequately compensated by established methods like feedforward control.

A more precise modelling of the machine and the inverter is required. It is preferably based on the use of complex state variables that accurately reflect the dynamics of the drive system. With such selection, an adequate controller emerges, having a transfer function with complex coefficients.

The high harmonic distortion due to the low switching frequency is a trade-off. The machine losses increase, as a consequence. Employing conventional carrier-based pulsewidth modulation methods for inverter control is not a viable solution [5].

The use of synchronous optimal pulsewidth modulation is a preferred alternative. The method optimizes the pulse patterns in an off-line procedure: it computes the switching angles over one fundamental period

3

for all possible operating points [6]. The resulting pulse patterns are such, that the harmonic distortion of the currents in the stator winding is restricted to minimum values.

Since the optimization of the pulse patterns is done off-line, steady-state operation must be assumed. The optimization is then not valid at changes of the operating point. Even at quasi steady-state operation, excursions of the harmonic currents will occur that may lead the drive to overcurrent conditions [7].

As an improvement, current trajectory tracking has been employed [8]. The method derives the optimal, steady-state stator current trajectory from the pulse pattern in use. The actual stator current vector is forced to follow this target trajectory. It is a disadvantage that the stator current trajectory depends on the parameters of the drive motor, specifically on the total leakage inductance [9]. Also changing load conditions have been found to influence the stator current trajectory.

In this work, the trajectory of the stator flux linkage vector is estimated and used as a tracking target [5]. Being insensitive to parameter variations, it is better suited for this purpose. A real-time optimization of the stator flux trajectory accounts for elimination of harmonic excursions that appear when the operating point changes.

Fast control of the machine torque requires the fundamental component of the stator current as a feedback signal. Such signal is inherently obtained as part of the modulation algorithm when space vector modulation is used [10]. Synchronous optimal modulation does not offer a comparable feature [9]. A method for identifying the instantaneous fundamental component of the stator current, and of the stator flux linkage vector, is proposed. A novel observer is developed for this purpose. Its performance is analyzed and compared with a classical full-order observer.

Control in closed-loop of the fundamental feedback signal is subsequently imposed. When operation at synchronous optimal modulation is assumed, the design concept cannot be based on conventional methods: these tend to interfere with the optimized pulse patterns, especially when larger changes of the operating point occur at transient operation.

The proposed method is therefore based on a nonlinear controller: it aims to achieve deadbeat behavior and complete decoupling, while preserving at the same time steady-state conditions required by synchronous optimal modulation [11].

4

2. Medium-voltage drives

Overview

Adjustable-speed drives control the angular velocity of the rotor shaft of ac electric machines by supplying them with voltage of variable frequency. In the high-power range, medium-voltage machines are employed in various industrial setups that demand adjustable frequency. These include cement and petrochemical applications, waste and sewage processing, mining, and forest product treatment.

The structure of medium-voltage drives comprises typically two states of power conversion:

- An input stage formed by a power rectifier of the three-phase supply voltage feeding the capacitors of a dc-link circuit, and
- an output stage consisting of a power inverter to transform the dc-link voltage of the capacitors into variable-frequency, switched voltage waveforms that are fed to the three-phase ac machine.

The block diagram in Fig. 2.1 shows the setup of a typical ac drive. The present work focuses on the control of the output stage of medium-voltage drives.

The three-phase induction machine is analyzed in Section 2.1; a mathematical model, based on complex state variables, is derived to describe the generation of electromagnetic torque and rotational speed.

Section 2.2 focuses on the three-level inverter that provides the machine with variable-frequency voltage. Emphasis is given to the neutral point clamped inverter topology, which is commonly employed in high-power industrial applications. The ac voltage waveforms produced by the inverter are described in terms of switching state vectors. The operating point of the ac drive is formalized.

Finally, a short description of the industrial implementation is included in Section 2.3: a 2.5-MVA inverter controlled by a DSP-based digital

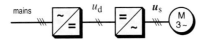

Fig. 2.1 Power conversion in an ac machine drive. The block diagram shows the input rectifier providing the dc-link voltage u_d and the power inverter supplying the induction machine with three-phase ac voltage $\boldsymbol{u}_s$ of adjustable frequency.

processing system is employed to experimentally verify the control methods that are detailed in the subsequent Chapters.

2.1 The three-phase induction machine

Induction machines are most commonly used in adjustable-speed ac drives. In the high-power range, induction machines are the workhorses of the industry. Motors of output power up to 0.5 MW are produced in standardized frame sizes allowing high interchangeability between manufacturers. Medium-voltage induction motors serve in setups of various types ranging from simple voltage-controlled industrial pumps and fans to more sophisticated drives requiring current control in closed-loop, such as in rolling mill, wind turbine, conveyor belt, traction and mobile offshore oil/gas drilling rigs applications.

2.1.1 Principle of operation

A three-phase induction machine of the squirrel-cage type is shown in Fig. 2.2(a). A two-pole machine is considered to simplify the investigation. Each phase of the motor is equipped with spatially distributed coils in both the stator and rotor windings. By supplying balanced three-phase ac voltage to the stator, a sinusoidal current density distribution is generated by its windings. It produces a spatially distributed flux density wave in the air gap that rotates at stator frequency ω_s; the air gap is assumed to be uniform along the periphery of the stator.

The rotor conductors are subjected to the magnetic field of the air gap. Current is induced in the short-circuited rotor windings. When observed from the rotor, the steady-state air gap flux revolves at slip frequency $\omega_r = \omega_s - \omega$, where ω is the angular velocity of the rotor shaft. The sinusoidal current density distribution in the rotor revolves also at slip frequency ω_r.

Tangential forces on the rotor surface result from the interaction between the flux density wave in the air gap and in the rotor winding; these forces produce electromagnetic torque T_e. Torque production is only possible if the angular velocity ω of the rotor differs from the stator frequency ω_s to allow the generation of rotor currents through induction.

2.1.2 The stator current space vector

Since the current density in the stator is sinusoidally distributed in space, it can be represented by the space vector A_s rotating at stator

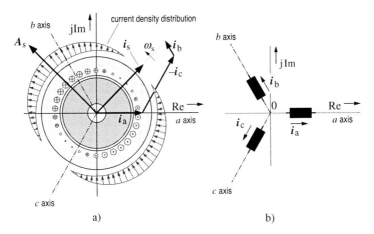

Fig. 2.2 Operating principle of the induction machine; (a) cross-section of the machine. The current density distribution A_s, and stator current space vector i_s are shown, (b) three-phase currents in the stator winding.

frequency ω_s [12]. The half-moon shapes in Fig. 2.2(a) provide a graphical interpretation of the current density wave; the space vector A_s is oriented towards the direction of its maximum intensity. It is generally preferred to use the current space vector i_s to describe the current density distribution [12]. The current space vector i_s is displaced by $-\pi/2$ with respect to A_s.

The scalar phase components i_a, i_b, and i_c, whose instantaneous values can be measured at the machine terminals, produce spatial current density distributions inside the machine. These sinusoidal distributions are represented by the respective space vectors i_a, i_b, and i_c, being displaced by $2\pi/3$ with respect to each other. They coincide with the directions of the respective winding axes, as shown in Fig. 2.2(b). The current space vector is obtained by superimposing the current density distributions of the three phases

$$i_s = \frac{2}{3}\left(1 \cdot i_a + ai_b + a^2 i_c\right), \qquad (2.1)$$

where $a = exp(j2\pi/3)$. The expression (2.1) is visualized in Fig. 2.2(a). The scalar phase components i_a, i_b, and i_c are obtained as projections of

7

the current space vector i_s on the respective phase axis

$$i_a = \Re\{i_s\}$$ (2.2a)

$$i_b = \Re\{a^2 \cdot i_s\}$$ (2.2b)

$$i_c = \Re\{a \cdot i_s\}.$$ (2.2c)

The stator current space vector in Fig. 2.2(a) is shown in the stationary coordinate system, being fixed to the stator, having its real axis aligned with phase a. The space vector i_s rotates at stator frequency ω_s with respect to the stationary coordinates.

2.1.3 System modelling by complex state variables

As with the stator current, a space vector can be also introduced to describe the sinusoidal distribution of the stator voltage inside the machine

$$u_s = r_s i_s + \frac{d\Psi_s}{d\tau}.$$ (2.3)

The quantities in (2.3) are normalized. To acquire the normalized machine current, the peak value of the rated phase current $\sqrt{2}I_{ph\,R}$ is chosen as base value. For the voltage variable, the peak value of the rated phase voltage $\sqrt{2}U_{ph\,R}$ is the base quantity. Time is also normalized as $\tau = \omega_{sR}t$, where ω_{sR} is the rated value of the stator frequency. The normalization of variables is explained in the Appendix.

The voltage space vectors $r_s i_s$ and $u_i = d\Psi_s/d\tau$ in (2.3) represent the voltage drop across the stator winding resistance r_s, and the induced voltage, respectively. The sinusoidal spatial distribution of the flux linkage with the stator winding is represented by the space vector Ψ_s in (2.3). The stator winding equation (2.3) is visualized in the vector diagram of Fig. 2.3. The space vector Ψ_r is also shown in Fig. 2.3: it describes the flux linkage distribution with the rotor winding.

The stator voltage expression (2.3) is referred to the stationary coordinate system (S). In a general case, a coordinate system rotating at arbitrary positive velocity ω_k may be selected to observe the complex variables of the induction machine. In the vector diagram of Fig. 2.3, the *field* coordinate system (F) is shown: it is in synchronism with the rotor flux space vector Ψ_r, having an angular velocity $\omega_k = \omega_s$ with respect to the stator. The field coordinate system is displaced by the rotor field angle δ with respect to the stationary coordinates.

Assuming constant stator frequency, $\omega_s = $ const., the field angle

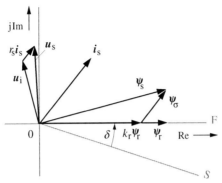

Fig. 2.3 Vector diagram showing the space vectors of the stator voltage, the flux linkage variables, and the stator current in an ac machine. The field coordinates (F) are displaced by the rotor field angle δ with respect to the stationary coordinate system (S).

$\delta = \omega_s \tau$. Transforming (2.3) from stationary (S) to field (F) coordinates is achieved by multiplying (2.3) by the unity vector rotator $exp(-j\omega_s \tau)$,

$$u_s = r_s i_s + \frac{d\psi_s}{d\tau} + j\omega_s \psi_s. \tag{2.4a}$$

This introduces the motion induced voltage $j\omega_s \psi_s$. It is contributed by the rotation of the stator winding with respect to the field coordinate system.

The rotor voltage equation in a squirrel-cage induction machine having short-circuited rotor windings, $u_r = 0$, is described in terms of space vectors

$$0 = r_r i_r + \frac{d\psi_r}{dt} + j(\omega_s - \omega)\psi_r, \tag{2.4b}$$

where r_r is the rotor resistance. The space vector i_r represents the current density distribution that is installed in the rotor winding by induction action.

The electrical subsystem of the induction machine is completed by considering the algebraic relationships between the space vectors of currents and flux linkages

$$\psi_s = l_s i_s + l_m i_r \tag{2.4c}$$

$$\psi_r = l_m i_s + l_r i_r. \tag{2.4d}$$

9

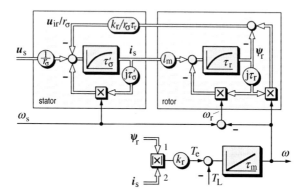

Fig. 2.4   Signal flow graph of the induction machine in field coordinates. The stator voltage $\boldsymbol{u}_s$, the angular velocity ω, and the rotational velocity of the field coordinate system ω_s are inputs to the electrical subsystem. The stator current $\boldsymbol{i}_s$ and rotor flux linkage $\boldsymbol{\Psi}_r$ are complex state variables. The lower portion of the graph shows the mechanical subsystem of the induction machine: the interaction of $\boldsymbol{i}_s$ and $\boldsymbol{\Psi}_r$ generates electromagnetic torque T_e which controls the motion of the rotor axis.

The respective parameters l_s and l_r in (2.4) represent the stator and the rotor inductance; l_m is the main inductance. The stator flux linkage vector $\boldsymbol{\Psi}_s$ shown in Fig. 2.3 is linked to the stator current vector by the relationship

$$\boldsymbol{\Psi}_s = k_r \boldsymbol{\Psi}_r + l_\sigma \boldsymbol{i}_s, \qquad (2.5)$$

which follows from (2.4). In (2.5), $k_r = l_m/l_r$ is the coupling factor of the rotor and $l_\sigma = \sigma l_s$ is the leakage inductance of the machine, where $\sigma = 1 - l_m^2/(l_s l_r)$ is the total leakage factor. The second term of the right-hand side of (2.5) represents the leakage flux $\boldsymbol{\Psi}_\sigma = l_\sigma \boldsymbol{i}_s$ of the machine. It is proportional to the stator current space vector $\boldsymbol{i}_s$.

The electrical subsystem of the induction machine (2.4) is of second order in terms of complex state variables. Considering the stator voltage $\boldsymbol{u}_s$, the angular mechanical velocity ω, and the rotational velocity of the field coordinate system ω_s as inputs, a machine model can be defined by selecting one suitable pair from the four state variables $\boldsymbol{i}_s$, $\boldsymbol{\Psi}_s$, $\boldsymbol{i}_r$, and $\boldsymbol{\Psi}_r$. The signal flow graph Fig. 2.4 represents the machine in terms of the stator current $\boldsymbol{i}_s$ and the rotor flux $\boldsymbol{\Psi}_r$ space vectors. The respec-

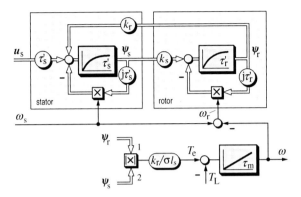

Fig. 2.5 Signal flow graph of the induction machine; the stator flux linkage Ψ_s and rotor flux linkage Ψ_r are selected as complex state variables.

tive equations in field coordinates are derived from (2.4):

$$\tau_\sigma' \frac{d i_s}{d\tau} + i_s = -j\omega_s \tau_\sigma' i_s - \frac{k_r}{\tau_r r_\sigma}(j\omega\tau_r - 1)\Psi_r + \frac{1}{r_\sigma} u_s \qquad (2.6a)$$

$$\tau_r \frac{d\Psi_r}{d\tau} + \Psi_r = -j\omega_r \tau_r \Psi_r + l_m i_s. \qquad (2.6b)$$

where $\tau_\sigma' = \sigma l_s / r_\sigma$ is a transient stator time constant and $\tau_r = l_r / r_r$ is the rotor time constant. The equivalent resistance r_σ is expressed by $r_\sigma = r_s + k_r^2 r_r$.

The electromagnetic torque T_e is computed as the z-component of the vector product of the rotor flux and stator current space vector

$$T_e = k_r |\Psi_r \times i_s|_z. \qquad (2.7)$$

The mechanical subsystem of the machine is modelled by the first-order equation of rotational motion

$$\tau_m \frac{d\omega}{dt} = T_e - T_L, \qquad (2.8)$$

from which the angular velocity ω of the rotor shaft is obtained. In (2.8), τ_m is the mechanical time constant of the machine and T_L is the variable load torque. The mechanical subsystem (2.8) is shown in the lower portion of the signal flow graph Fig. 2.4.

11

An equivalent definition of the machine model is given in the signal flow graph Fig. 2.5. The space vectors of the stator flux linkage $\boldsymbol{\psi}_s$ and of the rotor flux linkage $\boldsymbol{\psi}_r$ are selected as state variables. Equations (2.4) then result in

$$\tau_s' \frac{d\boldsymbol{\psi}_s}{d\tau} + \boldsymbol{\psi}_s = -j\omega_s\tau_s' \, \boldsymbol{\psi}_s + k_r\boldsymbol{\psi}_r + \tau_s' \boldsymbol{u}_s \qquad (2.9a)$$

$$\tau_r' \frac{d\boldsymbol{\psi}_r}{d\tau} + \boldsymbol{\psi}_r = -j\omega_r\tau_r' \boldsymbol{\psi}_r + k_s\boldsymbol{\psi}_s, \qquad (2.9b)$$

where $\tau_s' = \sigma l_s/r_s$ and $\tau_s' = \sigma l_r/r_r$ are the transient time constants of the stator and the rotor winding, respectively, and $k_s = l_m/l_s$ is the coupling factor of the stator. With such selection, the electromagnetic torque

$$T_e = \frac{k_r}{\sigma l_s} |\boldsymbol{\psi}_r \times \boldsymbol{\psi}_s|_z. \qquad (2.10)$$

2.1.4 Induction machine dynamics and cross-coupling

Observing the signal flow graph of Fig. 2.4, there exists a coupling from the stator to the rotor winding: it is expressed through the flux linkage space vector $l_m i_s$. This quantity acts as an input to the first-order structure representing the rotor. It expresses the contribution of the stator current distribution to the rotor flux linkage space vector. The rotor winding is characterized by the rotor time constant $\tau_r = l_r/r_r$ ranging from around 0.1 s (low-voltage machines) to 2.5 s (medium-voltage machines). Its relatively large value is owed to the large inductance l_r of the short-circuited rotor winding.

The stator winding dynamics are governed by the transient time constant τ_σ'. Its value ranges from less than 5 ms (low-voltage machines) to around 45 ms (medium-voltage machines). The rotor winding reacts on the stator winding through the rotor induced voltage

$$\boldsymbol{u}_{ir} = -\frac{k_r}{\tau_r}(j\omega\tau_r - 1)\boldsymbol{\psi}_r. \qquad (2.11)$$

The value of the rotor induced voltage depends mainly on the imaginary *cross-coupling* term $-j\omega\tau_r\boldsymbol{\psi}_r$, unless the angular velocity is very low. Another cross-coupling term exists through the internal feedback loop of the rotor winding: it is expressed by the space vector $-j\omega_r\tau_r\boldsymbol{\psi}_r$; its magnitude is proportional to the slip frequency ω_r.

The influence of these cross-coupling terms on the machine dynamics is negligible owing to the slow dynamic response of the rotor flux

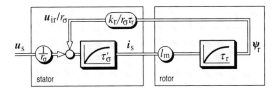

Fig. 2.6 Signal flow graph of the induction machine operating at zero speed, $\omega = 0$. The system is displayed in stationary coordinates, $\omega_k = 0$. All undesired cross-coupling terms are eliminated; the model reduces to a series connection of two first-order complex elements arranged in a positive-feedback loop.

linkage $\boldsymbol{\psi}_r$. An instantaneous change of the voltage space vector $\boldsymbol{u}_s$ at the input of the stator winding has an immediate effect on the stator current space vector $\boldsymbol{i}_s$; however, owing to the large value of the rotor time constant τ_r, a change of intensity and orientation of $\boldsymbol{i}_s$ does not significantly activate the cross-coupling of the rotor winding.

A cross-coupling term also exists through the motion-induced voltage $-j\omega_s l_\sigma \boldsymbol{i}_s$. This imaginary voltage appears when the stator equation (2.6a) is multiplied by the equivalent resistance r_σ. The motion-induced voltage is not an inherent quantity of the physical system: it depends upon the observation platform that is chosen to describe the machine dynamics.

This fact is confirmed by the signal flow graph Fig. 2.6, showing the machine model in stationary coordinates, $\omega_k = 0$. Observing the machine from the stationary coordinate system eliminates the motion induced voltage. In Fig. 2.6, the angular mechanical velocity is deliberately set $\omega = 0$. The system then reduces to a series connection of two linear first-order elements; all cross-coupling terms are eliminated.

It is obvious that the cross-coupling term $-j\omega_s l_\sigma \boldsymbol{i}_s$ cannot be omitted if the machine is modelled in field coordinates, as it describes the motion of the stator winding with respect to the field coordinate system. Since current controllers operating in closed-loop are typically arranged in field coordinates, the motion-induced voltage influences upon the dynamics of the system. One way to deal with cross-coupling is to apply feedforward compensation: this issue is examined in Section 4.6.

2.2 Voltage source inverters in medium-voltage drives

Medium-voltage drives are typically supplied by voltages starting at

1 kV; higher supply voltages are provided by standard 2.3 kV and 4.16 kV utilities (North and South American standards). In Europe and the rest of the world, standard supply voltage ratings are 3.3 kV and 6.6 kV. The output power generated by medium-voltage units starts at values lower than 1 MVA; state-of-the-art drives deliver shaft power between 2 MVA and 6 MVA. Output power ratings of over 15 MVA can be realized by customized designs using series or parallel connection of power semiconductor switches.

2.2.1 Power semiconductor switches

Power semiconductors of either the IGBT or IGCT type are typically used in modern high-power inverters. Both of them allow achieving high voltage and current ratings. The maximum repetitive peak forward blocking voltage of single IGCT switches is presently 6500 V at a maximum turn-off current rating of 4200 A [13]. IGBTs of maximum collector-emitter blocking voltage 6500 V at 600 A nominal collector current are available [13].

In high-power applications where neither series nor parallel connection of semiconductor devices is required, IGCTs are preferred candidate for inverter design: single switches guarantee maximum current ratings at high voltage. Compared to IGBTs of the same voltage rating, IGCTs offer the major advantage of considerably lower on-state and dynamic losses [14].

IGCTs require a turn-on snubber circuit to limit the rate of the current rise of the turning-off diodes in power converter configurations. On the other hand, IGBTs permit controlling the rate of the current rise and the rate of the voltage rise during switching transients by simply adjusting the gate drive unit. Compared to IGCTs, IGBTs require less amount of gate drive power for commutation purposes. Additional attractive features of the IGBT include the inherent limitation mechanism of short-circuit currents and the possibility of active clamping [14].

2.2.2 The two-level voltage source inverter

The circuit diagram Fig. 2.7(a) shows a conventional two-level IGBT voltage source inverter. A dc voltage supply u_d is provided at the inverter input by a low-impedance dc source. The dc-link voltage u_d is normalized; the quantity $(2/\pi)U_d$ is selected as base value resulting in $u_d = \pi/2$.

The inverter consists of three half-bridge units; the upper and lower power switches of each unit are alternatingly turned on and off at given

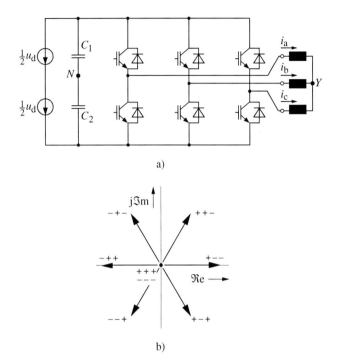

a)

b)

Fig. 2.7 Two-level voltage source inverter; (a) circuit diagram,
                                             (b) switching state vectors

time instants. Each of the three output terminals can be connected to either the positive dc-link voltage potential $+u_d/2$, or to the negative potential $-u_d/2$, depending on the state of the switches in the respective half-bridge. A total of $N_u = 2^3 = 8$ different arrangements of the switches are therefore possible. They are selected by the firing signals at the gates of the power semiconductors.

Each arrangement stands for a respective switching state vector u_k. The switching state vectors of the two-level inverter are shown in Fig. 2.7(b). The notation, e.g. $(+ - -)$, indicates that phase a is connected to the positive dc rail, whereas phases b and c are linked to the negative dc rail. Each switching state vector is characterized by its intensity and

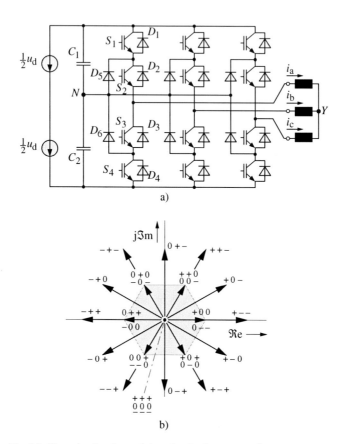

Fig. 2.8 Neutral point clamped three-level voltage source inverter;
(a) circuit diagram, (b) switching state vectors

orientation in the complex plane. These properties are derived from the
magnitude and phase angle of the respective vector

$$u_k = \frac{2}{3}\left(1 \cdot s_a + a s_b + a^2 s_c\right). \tag{2.12}$$

In (2.12), the variables s_a, s_b, and s_c denote the instantaneous value of
the selected state of each phase, $s_a, s_b, s_c \in \{-u_d/2, +u_d/2\}$.

2.2.3 The three-level neutral point clamped voltage source inverter topology

The circuit diagram Fig. 2.8(a) shows a three-level neutral point clamped inverter [2] being fed by a dc source of voltage u_d. The voltage of the dc-link circuit is split by the capacitors C_1 and C_2 into three levels. The common terminal point N between the capacitors serves as reference.

The output potential of each phase can acquire three discrete levels, $+u_d/2$, 0, or $-u_d/2$, depending on the control of the four power switches in the respective phase arm. Three different states are therefore identified for each phase; they are explained with reference to phase a:

- **Positive output potential**: If switches S_1 and S_2 are turned on, the phase-a potential $u_a = +u_d/2$. The voltage across the series connection of the deactivated switches S_3 and S_4 equals u_d. The clamping diode D_6 serves to sharing the dc-link voltage to equal values, i.e. $u_d/2$, across the power terminals of the switches S_3 and S_4. Positive load current flows in phase a from the positive rail of the dc-link through switches S_1 and S_2; a negative load current flows from the phase-a terminal back to the positive rail through the free-wheeling diodes D_1 and D_2.

- **Zero output potential**: If switches S_2 and S_3 are turned on, then $u_a = 0$. The clamping diodes D_5 and D_6 divide the dc-link voltage u_d, so as to maintain the voltage equilibrium across the turned-off switches S_1 and S_4. Positive current flows from the neutral point N to phase a through the clamping diode D_5 and switch S_2; negative current flows from phase a back to the neutral point N through the switch S_3 and the clamping diode D_6.

- **Negative output potential**: If switches S_3 and S_4 are turned on, then $u_a = -u_d/2$. The dc-link voltage u_d is equally shared between the two deactivated switches S_1 and S_2 by virtue of the clamping diode D_5. The phase current i_a flows in the positive direction through free-wheeling diodes D_4 and D_3. It flows back from phase a to the negative potential of the dc-link bus through the switches S_3 and S_4.

2.2.4 Advantages of the three-level inverter topology

The three-level inverter offers a number of attractive features which makes it a preferred candidate for medium-voltage applications:

- As compared with a two-level inverter that utilizes semiconductor switches of the same voltage and current rating, a three-level inverter allows operation at double the dc-link voltage.

- Reversing the previous argument, it is assumed that both topologies are fed by the same dc-link voltage u_d: in this case, semiconductor switches of lower voltage rating can be used in the three-level inverter. These allow shorter turn-on and turn-off times as compared to switches of higher voltage employed in two-level inverters; switching and conduction losses reduce as a consequence.
- Owing to the increased number of voltage potentials at its dc-link, the harmonic content of the output voltages and currents is reduced compared to a two-level inverter of the same dc-link voltage rating.
- The stress in the bearings of a motor driven by a three-level inverter is reduced owing to the low common-mode voltage [15].
- Finally, the rate of change du/dt of the output voltage in a three-level inverter is less, as the step amplitude is only $u_d/2$ as compared to u_d in a two-level inverter; the stress in the connection cables and in the motor windings is therefore lower.

In addition to the neutral point clamped inverter, there are more topologies of the three-level type, such as the flying-capacitor topology [16, 17], the cascaded multicell topology with separate dc sources [18], and the hybrid multilevel cell inverter [19]. The neutral point clamped voltage source inverter is meanwhile well established as an industrial standard in the high-power range [20], because of its simple and modular structure.

2.2.5 Switching state vectors of the three-level inverter

The switching state vectors u_k of the three-level inverter are visualized in the complex plane Fig. 2.8(b). There are a total of $N_u = 3^3 = 27$ switching state vectors for the three-level inverter by virtue of the third dc voltage potential for the connection of each phase. The magnitude and phase angle of each switching state vector is obtained by inserting the instantaneous values of the three voltage levels s_a, s_b, $s_c \in \{-u_d/2,$ $0, +u_d/2\}$ in (2.12). The vectors u_k can be classified in four groups:

1. Three zero vectors, (000), (+++), and (– – –), generate zero voltage at the inverter output, since the machine phases are short-circuited, being all connected to one of the dc-link rails.
2. Medium vectors, such as (+ 0 –) or (0 – +), connect exactly one of the three phase terminals to the neutral point, whereas the other two phase terminals are connected to the positive and negative dc-rails.

3. Large vectors, such as (+ – –) or (+ + –), connect the phase terminals either to the positive or the negative dc rail.

4. Small vectors are categorized in two subgroups: *positive* small vectors, such as (+ 0 0) or (+ + 0), result when at least one inverter phase is connected to the positive dc rail, whereas the remaining phases are linked to the neutral point N. If at least one phase terminal is connected to the negative dc rail, and the remaining phases to the common terminal point N, then *negative* small vectors are generated, such as (0 – –) or (0 0 –). The small vectors of the two subgroups form *redundant* vector pairs, e.g. (+ 0 0) and (0 – –). Redundant switching state vectors have equal magnitude and phase angle according to (2.12). Fig. 2.8(b) shows that 6 pairs of redundant switching state vectors exist in a three-level inverter.

Taking into account the redundancy of the small vectors, there are 18 unique active switching vectors in a three-level inverter producing nonzero output voltage.

2.2.6 Three-phase switched voltage waveforms

The switching state vectors u_k of Fig. 2.8(b) are selected by employing *pulsewidth modulation* methods. These generate three-phase, switched voltage waveforms by varying the number and width of voltage pulses depending on the operating point. A example of pulsewidth-modulated voltage waveforms is given in Fig. 2.9. The three-phase output potential waveforms u_a, u_b, $u_c \in \{-u_d/2, 0, +u_d/2\}$ in Fig. 2.9(a) feed an ac machine whose windings are arranged in star connection as in the circuit diagram of Fig. 2.8(a). The fundamental frequency of the voltage waveforms $f_1 = 45$ Hz. The balanced switched voltages in Fig. 2.9(a) are displaced by $2\pi/3$ with respect to each other over one fundamental period T_1.

Fig. 2.9(b) displays the potential of the common connection terminal Y of the machine phases in Fig. 2.8(a)

$$u_Y = \frac{1}{3}(u_a + u_b + u_c). \tag{2.13}$$

Its maximum amplitude amounts to one-third of the dc-link voltage u_d. Its frequency is three times the frequency f_1 of the phase potentials. Consequently, the waveform of the neutral point potential u_Y in Fig. 2.9(b) includes all triplen harmonics of the phase potentials u_a, u_b, and u_c. The waveform of the phase-a voltage,

$$u_{sa} = u_a - u_Y, \tag{2.14}$$

19

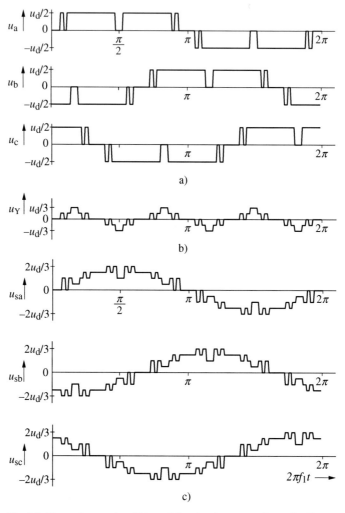

Fig. 2.9 Three-phase pulsewidth-modulated voltage waveforms feeding a three-level voltage source inventer at $f_1 = 45$Hz, (a) switched phase potentials u_a, u_b, and u_c, (b) potential of the common connection terminal of the machine phases u_Y, (c) three-phase switched voltage waveforms u_{sa}, u_{sb}, and u_{sc}

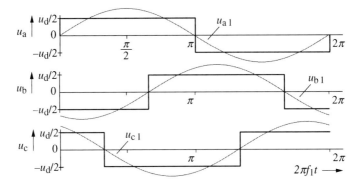

Fig. 2.10 Switched phase potentials of a three-phase ac system operating at six-step mode; the fundamental frequency $f_1 = 50$ Hz. The sinusoidal waveforms show the fundamental component of the potential waveforms.

is therefore free of triplen harmonics; this applies also for u_{sb} and u_{sc}.

The phase voltage waveforms u_{sa}, u_{sb}, and u_{sc} are displayed in Fig. 2.9(c). By inspecting expressions (2.13) and (2.14), it is evident that the instantaneous value of phase voltages u_{sa} does not only depend upon the potential u_a: the other two-phase potentials u_b and u_c also contribute to its value. This is generally true in any three-phase system fed by non-sinusoidal voltage waveforms generated by a pulsewidth-modulated inverter.

The instantaneous values u_{sa}, u_{sb}, and u_{sc} of the respective phase voltages are introduced in (2.12):

$$u_s = \frac{2}{3}\left(1 \cdot u_{sa} + au_{sb} + a^2 u_{sc}\right) \qquad (2.15)$$

to define the space vector of the phase voltage that is generated at the output terminals of the inverter.

The fundamental voltage magnitude of the modulated switching sequence is obtained with reference to the Fourier Series analysis of u_a as

$$u_1 = \frac{4}{T_1} \int_0^{T_1} u_a(t)\, sin(2\pi f_1 t)\, dt . \qquad (2.16)$$

The maximum achievable fundamental voltage value is attained at operation in the *six-step mode* which is visualized in Fig. 2.10 for

21

operation at a fundamental frequency f_1 = 50 Hz. In the six-step mode, each phase is connected to the positive potential of the dc-link circuit of the inverter for one half of the fundamental period, i.e. for $T_1/2$ = $1/(2f_1)$; the respective phases are then connected to the negative potential during the next half-wave. The fundamental voltage $u_{1\ \text{six-step}}$ = $(\pi/4)u_d$.

Having discussed the properties of three-phase switched voltage waveforms, it is possible to formalize the definition of the operating point of the system: it is characterized by the fundamental frequency f_1, and by the modulation index m. The latter quantity is the normalized fundamental voltage at the inverter output

$$m = \frac{u_1}{u_{1\ \text{six-step}}}. \tag{2.17}$$

The switching frequency is the number of commutations N of each power semiconductor switch within a fundamental period T_1, multiplied by the respective fundamental frequency f_1

$$f_s = N \cdot f_1. \tag{2.18}$$

Exemplifying (2.18), the stationary three-phase voltage waveforms of Fig. 2.9 correspond to operation at switching frequency $f_s = 4 \cdot 45 = 180$ Hz, whereas the switching frequency at six-step mode, Fig. 2.10, is $f_s = 1 \cdot 50 = 50$ Hz.

2.3   Industrial implementation

A 2.5-MVA, 300-A high-power inverter fed by 4.16-kV 60-Hz three-phase mains, served for the intents of the present work. The three-level inverter power circuit uses 6.5-kV, 300-A *EUPEC* IGBTs. It feeds a 1800-kW, 300-A medium-voltage machine; the machine parameter are listed in Table 1 of the Appendix. The medium-voltage ac drive is installed in an industrial environment.

A 36-kVA, 63-A three-level inverter prototype fed by 380-V, 50-Hz three-phase mains was employed for software development and testing purposes. This lower-power inverter employs 1200-V, 50-A *EUPEC* power switches of the IGBT type. A 30-kW, 60-A induction machine is fed by the low-power inverter; the machine parameter data are listed in Table 2 of the Appendix.

The digital processing system of both the medium-voltage drive and low-voltage prototype is based on the 32-bit *Hitachi* SH-4 7750 DSP operating at 167 MHz clock frequency. The PWM signals are generated by the microprocessor and processed in CPLDs of *Xilinx* to inject the lockout time between the gate pulses. The lockout time interval is set to 20 μs also in the low-power prototype to match the commutation requirements of medium voltage IGBTs. The firing signals are fed to *Agilent* gate drivers via optic fibers and an optocoupler circuit.

Further features of the electronics system include three-phase current measurement by *LEM* current-to-voltage transducers and a 10-bit A/D-conversion bus. The dc-link voltage is acquired by a voltage-to-frequency converter circuit. The terminal voltages of the machine are additional feedback signals: they are processed by *Texas Instruments* differential line receivers and converted to digital signals to be input to the microprocessor. The angular mechanical velocity of the machine is acquired by differentiating the rotor position signal, which is measured by a 1024-line encoder. Protection units based on analog operational amplifier circuits are implemented to avoid overcurrent and overvoltage conditions.

3. Medium-voltage drives operating at space vector modulation

Overview

Carrier-based modulation methods are most frequently used for the control of voltage source inverters to allow operation at variable frequency [21]. They generate switched voltage waveforms that depend on the operating point of the system.The dc power accumulated in the dc-link circuit of the inverter is then converted to three-phase ac power. Following the definition of the modulation law in Section 3.1, a common implementation of carrier modulation – the space vector modulation method – is explained in Section 3.1.1.

Section 3.2 introduces an important issue concerning the performance of pulsewidth-modulated drives: the variation of the neutral point potential in three-level NPC–inverter systems. The redundancy of switching state vectors available in three- level inverters is exploited to allow zero average neutral point potential values. High magnitude errors invoked at dynamic operation are compensated in closed-loop by selecting appropriate space vector modulation switching sequences. The method is discussed in Section 3.3.

3.1 Carrier-based pulsewidth modulation

There exists a number of different implementations of carrier-based modulation schemes, such as the suboscillation method and its modified variants, synchronized carrier modulation, and space vector modulation [22]. Despite the difference between carrier modulation schemes in terms of the involved implementation algorithms, and in terms of their stationary and dynamic performance, they all share two common characteristics:

- They are implemented in open-loop. They generate three-phase switched voltage waveforms whose fundamental space vector u_1 equals a given reference value

$$u_1(t_s) = u^*(t_s). \qquad (3.1)$$

Here, $u^*(t_s)$ is the voltage reference vector that serves as the input to the open-loop modulation algorithm, and t_s is the time instant at which this voltage is sampled. Open-loop modulation schemes are implemented with reference to Fig. 3.1. The reference voltage space

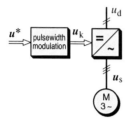

Fig. 3.1 Open-loop modulation scheme, signal flow graph

vector u^* is generated by superimposed controllers of the machine torque and excitation current.

• Carrier modulation schemes allocate the commutation instants of the inverter semiconductor devices within a *fixed* time interval, the subcycle $T_0 = 1/(2f_s)$, where f_s is the switching frequency [10, 21]. In the case of a three-level inverter system, the twelve commutations of the twelve-pulse inverter are generated periodically within a subcycle T_0 to form the respective switching state vectors u_k.

3.1.1 Space vector modulation

The signal flow graph in Fig. 3.2 visualizes the implementation process of space vector modulation [23]. The reference voltage vector u^* is sampled at double the switching frequency f_s. The application of the modulation law (3.1) takes place within the time interval $T_0 = 1/(2f_s)$. It is evaluated by expressions

$$\int_{(t_a)} u_a dt + \int_{(t_b)} u_b dt + \int_{(t_c)} u_c dt = \int_{(T_0)} u^* dt \tag{3.2a}$$

$$T_0 = t_a + t_b + t_c, \tag{3.2b}$$

which equate the volt-second value of the reference voltage sample $u^*(t_s)$ to that of a switching sequence $u_a\langle t_a\rangle \cdots u_b\langle t_b\rangle \cdots u_c\langle t_c\rangle$ of a subcycle T_0. The switching state vectors u_a, u_b, u_c are located in closest distance from the reference vector u^*. The respective on-durations t_a, t_b, t_c are associated to the switching state vectors u_a, u_b, u_c. They are obtained from the solution of (3.2).

An example is given in the diagram Fig. 3.3. Referring to a three-level inverter, operation in the low modulation index range is assumed. In this range, only *small* switching state vectors $u_1\langle t_1\rangle$ and $u_2\langle t_2\rangle$ and the zero vector $u_0\langle t_0\rangle$ form part of the switching sequences. Possible sequences in time of the switching state vectors are

$$(\text{H}^+) \equiv u_0\langle t_0/2\rangle \cdots u_a\langle t_a\rangle \cdots u_b\langle t_b\rangle \cdots u_0\langle t_0/2\rangle \ \in u_k \tag{3.3a}$$

$$(\text{H}^-) \equiv u_0\langle t_0/2\rangle \cdots u_b\langle t_b\rangle \cdots u_a\langle t_a\rangle \cdots u_0\langle t_0/2\rangle \ \in u_k \tag{3.3b}$$

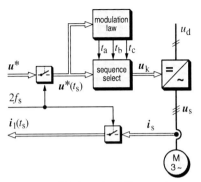

Fig. 3.2   Space vector modulation, signal flow graph

which defines the classical space vector modulation [10]. The sequence (3.3) consists of six switching states within an interval $2T_0$. The switching state vectors of the negative subcycle (H$^-$) in (3.3b) are the same as of the positive subcycle (H$^+$) in (3.3a), but arranged in a reversed sequence.

The oscillogram in Fig. 3.4 was recorded using the medium-voltage machine fed by the industrial three-level inverter; the setup was described in Section 2.3. The modulation index $m = 0.67$ and the fundamental frequency $f_1 = 33.3$ Hz. Fig. 3.4(a) shows the a-phase potential waveform generated by space vector modulation. The respective a-phase current i_a is given in Fig. 3.4(b); the dashed sinusoidal waveform is the fundamental current i_{a1}. In the trace Fig. 3.4(c), the a-phase harmonic current i_{ah} is shown; notice that the current scales in Fig. 3.4(b) and (c) are different.

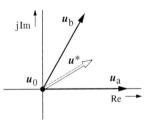

Fig. 3.3   Reference voltage space vector u^* and its neighboring switching state vectors u_0, u_1, and u_2

The harmonic current space vector is defined

$$i_h = i_s - i_1, \qquad (3.4)$$

where i_s is the stator current vector and i_1 is its fundamental component. The symmetry of the sequence (3.3) ensures that the trajectory of the harmonic current describes a closed pattern centered in the origin of the

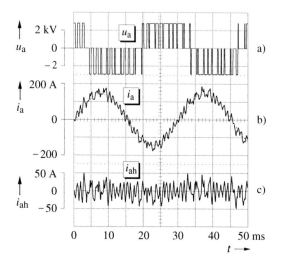

Fig. 3.4 Operation at space vector modulation at $m = 0.67$ and $f_1 = 33.3$ Hz in a three-level medium-voltage inverter; (a) phase potential u_a, (b) phase current i_a, (c) harmonic current i_{ah}; all waveforms refer to phase a.

complex plane, as shown in Fig. 3.5 at 200 Hz switching frequency and modulation index $m = 0.85$.

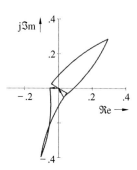

Fig. 3.5 Space vector modulation; trajectory of the harmonic current i_h, plotted over two half-cycles, $m = 0.85$ at switching frequency $f_s = 200$ Hz

During the on-duration of a zero vector u_0, the trajectory i_h of the harmonic current passes through the origin of the complex plane, Fig. 3.5. Given the symmetry of its track, the harmonic current equals zero when T_0 has elapsed, i.e. at the end of each subcycle. Hence $i_h(t_s) = 0$, where t_s is the time instant at which the reference voltage vector is sampled. We have therefore from (3.4), $i_s(t_s) = i_1(t_s)$ which permits measuring the fundamental current at the respective sampling instant t_s.

28

The signal flow graph Fig. 3.2 indicates how the fundamental current is acquired.

3.2 The neutral point potential problem

The circuit diagram Fig. 2.8(a) of the three-level inverter is repeated in Fig. 3.6. It shows that the two capacitors C_1 and C_2 of the dc-link serve as a voltage divider. The capacitor voltages u_{d1} and u_{d2} are generally not stabilized by external sources.

3.2.1 Effect of switching state vectors on the neutral point potential and current

The value of the neutral point potential changes in proportion to the integral of the neutral point current i_n, Fig. 3.6. Apart from the load conditions, it is the respective switching state of the inverter that determines the magnitude and sign of the neutral point current.

The effect of the switching state vectors u_k on the neutral point current and voltage is explained here by taking into consideration their classification in four groups, Section 2.2.5. A replica of the vector diagram Fig. 2.8(b) is provided in Fig. 3.7 to assist the reader.

1. The zero vectors $u_0^+ = u_0^{(+ + +)}$, $u_0^0 = u_0^{(0\,0\,0)}$ and $u_0^- = u_0^{(- - -)}$ do not cause a current flow through the neutral point N. Hence the neutral point potential is not affected.

2. Large vectors, such as $(+ - -)$, do not connect any of the phase terminals to the neutral point potential of the dc-link. They do not

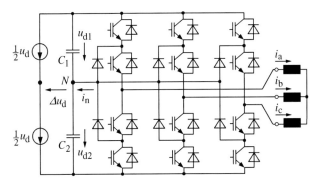

Fig. 3.6   Circuit diagram of a three-level inverter, NPC topology

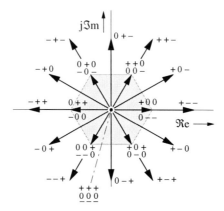

Fig. 3.7 Switching state vectors of a three-level inverter; the lettering indicates the respective combination of the three phase potentials.

cause a current flow through the neutral point which also leaves the neutral point potential unaffected.

3. Medium vectors, such as $(+ 0 -)$, generate current in the neutral point N, since they connect one of the three phase terminals to this point. The neutral point current changes then the neutral point potential.

4. Small vectors exist in redundant pairs: Each positive small vector, such as $u_1^+ = u_1^{(+\,0\,0)}$ has a corresponding negative small vector $u_1^- = u_1^{(0\,-\,-)}$ that generates the same output voltage vector. Although the neutral point current has the same magnitude in either case, its sign reverses depending on which of the two redundant vectors is activated. The neutral point potential then changes either to higher or to lower values.

It is therefore the small and the medium switching state vectors that introduce unbalances of the dc-link capacitor voltages. Such unbalances can be expressed by the neutral point potential error

$$\Delta u_n = u_{d1} - u_{d2}. \tag{3.5}$$

Neutral point potential errors can be excessive in medium voltage inverters since the dc link capacitors are normally not of the electrolytic type. Metal film capacitors are used instead; they can withstand higher

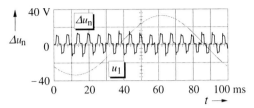

Fig. 3.8 Measured waveform of the neutral point potential error at 5.8 kV dc-link voltage; space vector modulation at low modulation index, f_s = 330 Hz switching frequency, f_1 = 10 Hz fundamental frequency.

rms harmonic currents per microfarad. Their effective capacitance is therefore much smaller than that of the electrolytic capacitors used in low voltage inverters. The resulting high neutral point potential errors of medium voltage inverters impose higher voltage stress on the semiconductor devices, and on the capacitors. The high voltage excursions amplify another effect: If a neutral point potential error exists, the actual voltages at the machine terminals are either higher or lower than their respective command values whenever medium or small switching state vectors are applied. The intended voltseconds of the pulsewidth modulation are then in error, which also changes the machine currents, and consequently the neutral point current i_n.

3.2.2 Steady-state neutral point potential error

A nonzero neutral point current i_n exists whenever either a small vector or a medium vector forms part of the switching sequence. The neutral point potential changes in consequence. During steady-state operation, a ripple appears on the neutral point potential waveform. Operating at low switching frequency increases all on-state durations. The ripple amplitudes are therefore high in medium-voltage applications.

The waveform of the neutral point ripple also depends on the modulation strategy. The frequency of the neutral point ripple equals the *switching frequency* at carrier based modulation. A measured waveform is shown in Fig. 3.8 as an example. A further oscillographed waveform is given in Fig. 3.9. In this case, the drive is operated at optimal pulsewidth modulation based on precalculated steady-state patterns; the method is explained in detail in Chapter 5. Fig. 3.9 illustrates that it produces a dominant neutral point ripple of *three times the fundamental frequency.*

31

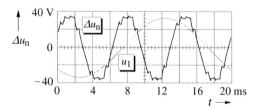

Fig. 3.9 Measured waveform of the neutral point potential error at 5.8 kV dc-link voltage; synchronous optimal modulation at high modulation index, f_s = 200 Hz switching frequency, f_1 = 50 Hz fundamental frequency.

3.2.3 Natural balancing

Three-level inverters have an intrinsic natural balancing mechanism that gives the average neutral point potential error the tendency to disappear [3]. The reason for this phenomenon is that an existing unbalance creates higher current changes in the positive direction when the neutral point potential error is positive, i.e. $u_{d1} > u_{d2}$ in Fig. 3.6, and vice versa. These conditions hold if the modulation operates at steady-state, not exerting an effect on the average zero error.

The natural balancing mechanism thus ensures the gradual elimination of the error at steady-state operation. An experimental proof is given by the oscillogram Fig. 3.10. A neutral point potential error generated by a transient at t_1 decays to almost zero within about 1.2 s.

3.2.4 Transient neutral point potential error

When the modulation index changes at transient operation, excursions of the neutral point potential may occur. Particularly during intervals of

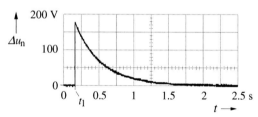

Fig. 3.10 Measured waveform illustrating the gradual elimination of a neutral point error by virtue of the natural balancing property; switching frequency f_s = 330 Hz, modulation index m = 0.25.

32

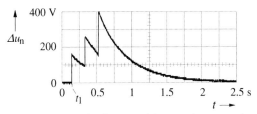

Fig. 3.11 Multiple neutral point potential errors due to transient operation and gradual elimination of a neutral point error by virtue of the natural balancing property; switching frequency 330 Hz, modulation index 0.25.

high dynamic operation will frequent changes of the modulation index lead to multiple neutral point potential errors within a limited time. Since the neutral point potential error decays only slowly by virtue of the natural balancing mechanism, the errors may accumulate up to the extreme of an overvoltage condition. Such process is illustrated in the oscillogram Fig. 3.11. Note that the amplitude scales in Fig. 3.10 and Fig. 3.11 are different. The examples demonstrate that fast neutral point potential control is required to eliminate neutral point potential errors at transient operation.

3.3 Control of the neutral point potential

3.3.1 State-of-the-art

Modifications of the pulse pattern must be done when a neutral point potential error exists. A first method exchanges one of the small switching state vectors by its redundant equivalent [24]. This technique may require multiple commutations between two consecutive switching states. To give an example, an original sequence

$$u_1^{(+\,0\,0)} \cdots u_0^{(0\,0\,0)} \cdots u_2^{(0\,0\,-)} \cdots$$
$$\cdots u_0^{(0\,0\,0)} \cdots u_1^{(+\,0\,0)}$$

(3.6a)

is modified by replacing $u_2^{(0\,0\,-)}$ with $u_2^{(+\,+\,0)}$ to yield

$$u_1^{(+\,0\,0)} \cdots u_0^{(0\,0\,0)} \cdots u_2^{(+\,+\,0)} \cdots$$
$$\cdots u_0^{(0\,0\,0)} \cdots u_1^{(+\,0\,0)}.$$

(3.6b)

This introduces two commutations in addition to the four commutations of the original sequence. The modification thus increases the switching losses which is particularly undesirable in high-power inverters.

A second method to control the neutral point potential error is presented in [25]. Depending on the direction of the neutral point current i_n, a zero sequence voltage component is either added to, or substracted from, the phase reference voltages. The neutral point current is thereby inverted, if required, in which process additional commutations may occur. The method was tested for small neutral point potential errors, up to 1.5% of the dc link voltage. It has shown to give good results in this case [25].

Acquiring the direction of the neutral point current has proved to be problematic when high current distortions exist at low switching frequency [26].

A third method controls the on-durations of a pair of redundant switching state vectors [21]. As an example, the space vector sequence

$$u_1^{(0--)}\langle t_1^-\rangle \cdots u_2^{(0\,0\,-)}\langle t_2^-\rangle \cdots u_0^{(0\,0\,0)}\langle t_0\rangle \cdots u_1^{(+\,0\,0)}\langle 2t_1^+\rangle$$

$$\cdots u_0^{(0\,0\,0)}\langle t_0\rangle \cdots u_2^{(0\,0\,-)}\langle t_2^-\rangle \cdots u_1^{(0--)}\langle t_1^-\rangle \qquad (3.7)$$

contains the redundant pair $u_1^{(0--)}\|u_1^{(+\,0\,0)}$. These two switching state vectors change the neutral point potential in opposed directions. Their original on-durations are both t_1. Modifying the durations as $t_1^- = t_1 + \Delta t$ and $t_1^+ = t_1 - \Delta t$, respectively, maintains the fundamental volt-seconds of the complete cycle (3.7), while introducing a change of the neutral point potential which is proportional to Δt.

The resulting trajectories of the harmonic current $i_h = i_s - i_1$ are exemplified in Fig. 3.12, where i_s is the stator current vector of an induction motor and i_1 is its fundamental component. Fig. 3.12(a) displays the switching state vectors and the corresponding current derivative vectors, approximated as $di_s(u_k)/dt = (u_k - u^*)/l_\sigma$, where u_k stands for any one of the switching state vectors and is l_σ the total leakage inductance of the motor. It is further assumed that the induced voltage of the motor $u_i = u^*$. The trajectory of the harmonic current vector of an unmodified switching sequence in Fig. 3.12(b) forms a closed pattern, centered in the origin of the complex plane. Against this, the modified switching sequence Fig. 3.12(c) has the on-duration of u_1^- extended to $t_1^- = t_1 + \Delta t$ while the on-duration of u_1^+ is reduced to $t_1^+ = t_1 - \Delta t$ by the same value Δt. It is obvious from the graph that the total harmonic distortion of the modified sequence is larger. Moreover, the maximum possible time extension is $\Delta t_{max} = 2t_1 - t_{min}$, where t_{min} is the mini-

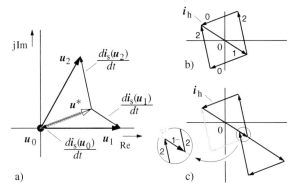

a)

b)

c)

Fig. 3.12 Effect of modified on-durations; (a) switching state vectors and current derivatives, (b) unmodified harmonic trajectory, (c) modified harmonic trajectory

mum on-duration of the semiconductor devices. If t_1 happens to be small in a particular situation, the compensation has little effect.

Using this method, the oscillogram Fig. 3.13 shows the compensation of an initial neutral point potential error $\Delta u_n = 400$ V, which is 7%

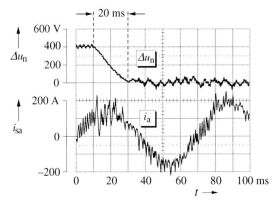

Fig. 3.13 Measured waveforms illustrating the performance of neutral point potential balancing by control of the on-time durations; switching frequency 330 Hz, modulation index 0.25; upper trace: neutral point potential error; lower trace: line current of phase a.

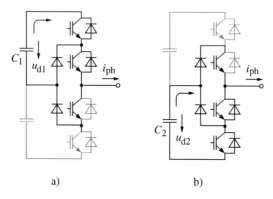

a)   b)

Fig. 3.14 Circuit topology of one bridge arm; (a) positive subbridge configuration, (b) negative subbridge configuration.

of the 5.8 kV dc link voltage in the 2.5-MVA medium-voltage inverter
of the industrial implementation, Section 2.3. The increased harmonic
distortion of the phase current during the compensation process is clearly
visible.

It is concluded that state-of-the-art compensation methods for neutral point potential error either create additional commutations that increase the switching losses, or, alternatively, interfere with the harmonic volt-second balance commanded by the pulsewidth modulator, thus
increasing the harmonic distortion.

3.3.2 Neutral point potential control by subbridge selection

The proposed method assumes operation at space vector modulation in
the low modulation index range. The upper range, where medium and
large switching state vectors are used, is not addressed in this work: it is
only the area within the shaded hexagon of Fig. 3.7 where space vector
modulation is of practical concern in industrial medium-voltage drives.
At higher modulation index values, space vector modulation is not preferred as it produces currents of high harmonic amplitudes in the machine windings; the related issues are detailed in Sections 5.1 and 5.4.

The approach described in the present Section achieves control of the
neutral point potential by exploiting the redundancy of switching state
vectors.

3.3.2.1 The concept of redundant subbridges

In the lower modulation index range, only small and zero vectors are used. Switching sequences can be then defined that activate exclusively one of the two *subbridge* configurations shown in Fig. 3.14. The sequences are

$$(\text{S}^+) \qquad u_0^0 \langle t_0/2 \rangle \cdots u_1^+ \langle t_1 \rangle \cdots u_2^+ \langle t_2 \rangle \cdots u_0^+ \langle t_0 \rangle \cdots$$
$$\cdots u_2^+ \langle t_2 \rangle \cdots u_1^+ \langle t_1 \rangle \cdots u_0^0 \langle t_0/2 \rangle \qquad (3.8a)$$

which uses only the semiconductor devices of the positive subbridge (S^+), and

$$(\text{S}^-) \qquad u_0^0 \langle t_0/2 \rangle \cdots u_2^- \langle t_2 \rangle \cdots u_1^- \langle t_1 \rangle \cdots u_0^- \langle t_0 \rangle \cdots$$
$$\cdots u_1^- \langle t_1 \rangle \cdots u_2^- \langle t_2 \rangle \cdots u_0^0 \langle t_0/2 \rangle \qquad (3.8b)$$

which uses only the devices of the negative subbridge (S^-).

Of interest is the polarity of the neutral point current in either case. The switching state of the inverter bridge determines how the phase currents contribute to the neutral point current. For example, if the positive subbridge is active selecting the switching state vector $(+\ 0\ 0)$, then $i_\text{n}(t) = i_\text{b}(t) + i_\text{c}(t) = -i_\text{a}(t)$. Similarly, if the negative subbridge is active selecting vector $(0\ -\ -)$, we have $i_\text{n}(t) = i_\text{a}(t)$. Zero vectors do not contribute to the neutral point current.

When the positive subbridge is active, the average neutral point current over a PWM cycle amounts to

$$\bar{i}_\text{n}^+ = \frac{1}{2T_0}[t_1(-i_\text{a}) + t_2 i_\text{c} + t_2 i_\text{c} + t_1(-i_\text{a})] = \frac{1}{T_0}(t_2 i_\text{c} - t_1 i_\text{a}) \quad (3.9a)$$

With the negative subbridge being active, the average neutral point current is

$$\bar{i}_\text{n}^- = \frac{1}{2T_0}[t_2(-i_\text{c}) + t_1 i_\text{a} + t_1 i_\text{a} + t_2(-i_\text{c})] = \frac{1}{T_0}(t_1 i_\text{a} - t_2 i_\text{c}) \quad (3.9b)$$

We obtain from equations (3.9a) and (3.9b)

$$\bar{i}_\text{n}^+ = -\bar{i}_\text{n}^- \qquad (3.10)$$

which means that the neutral point potential at steady-state is balanced if sequences (S^+) from (3.9a) and (S^-) from (3.9b) are alternatingly used. As both sequences start and terminate with the same switching state $u_0^{(0\,0\,0)}$, no additional commutations are required at the transitions. The steady-state waveform of the neutral point potential error Fig. 3.8 shows the expected alternating excursions.

Also of influence on the neutral point potential is the load condition of the machine. The space vector diagram Fig. 3.15 is referred to for clarification. To simplify the discussion, the reference voltage vector is assumed as $u^* = u^* exp(j\,30°)$ which results in $t_1 = t_2$. Ideal no-load operation is assumed for the machine, hence $i_s = i_s\,exp(-j\,60°)$. The average neutral point current over one PWM cycle with the positive subbridge being active, $\bar{i}_n{}^+ = (t_1/T_0) \cdot i_s\,(\cos 60° - \cos 60°) = 0$, which follows from (3.9a). Hence the neutral point potential does not change.

It is now assumed that light load is applied, for instance only the friction torque of the machine. The stator current space vector i_s is then displaced by a small value $\Delta\alpha$ in the positive direction. The average neutral point current is then proportional to $\cos\{(60° + \Delta\alpha) - \cos(60° - \Delta\alpha)\} = -\sqrt{3}\sin\Delta\alpha$ and hence becomes negative. At generating operation, the current displacement angle becomes negative – the average neutral point current is then positive.

The discussion shows that the average neutral point current is zero at no-load; it increases to negative values at motoring and to positive values at generating operation when the positive subbridge is used for modulation. According to (3.9b), conditions change when the negative subbridge is active. The average neutral point current is then negative at motoring and positive at generating operation.

3.3.2.2 Closed-loop control of the neutral point potential

Although the steady-state waveform of the neutral point potential exhibits a ripple content, Fig. 3.8, its average value is zero. Neutral point potential errors may occur during transients; errors of repeated transient conditions can successively add up and hence lead to high excursions, Fig. 3.11. It is important, therefore, that closed-loop control of the neutral point potential restores the zero average error whenever the error exceeds the amplitude of the steady-state ripple. Excessive ov-

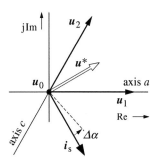

Fig. 3.15 Space vector diagram illustrating the effect of load flow on the neutral point current.

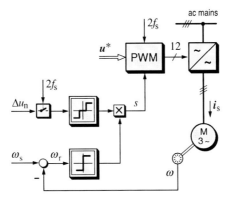

Fig. 3.16 Signal flow graph of the pulsewith modulator and neutral point potential controller.

ervoltage due to the accumulation of repetitive transient errors can be only reduced if the control acts fast.

The control is activated when the neutral point potential goes beyond the limits of a tolerance band, which is set to about double the neutral point potential ripple amplitude at steady-state.

A signal flow graph of the neutral point potential control scheme is shown in Fig. 3.16. The neutral point potential error Δu_n is sampled at frequency $2f_s$ in synchronism with the space vector modulation algorithm. If the tolerance value is exceeded, the algorithm interferes with the periodic steady-state changes between the switching sequences (S^+) and (S^-). Depending on whether the drive operates in the motoring or the generating mode, that particular sequence is selected that reduces the neutral point potential error. The decision is based on the sign of the electrical angular rotor frequency ω_r, or slip signal, which is readily available from the superimposed drive control system. The resulting sub-bridge selection signal s assumes the following values:

$s = 1$ selects the positive subbridge
$s = -1$ selects the negative subbridge
$s = 0$ deactivates the control, subbridges alternate.

The condition $s = 0$ indicates that the neutral point potential has returned to the voltage tolerance band. Steady-state operation of the modulator is then resumed.

39

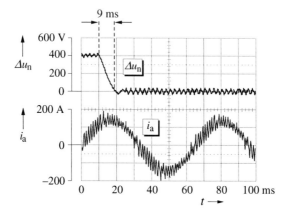

Fig. 3.17 Measured waveforms illustrating the transient behavior of neutral point balancing by redundant subbridge control; switching frequency 330 Hz, modulation index 0.25; upper trace: neutral point potential error; lower trace: line current waveform of phase a.

Relying on the rotor frequency signal ω_r for subbridge selection ensures a fast reaction of the neutral point potential control system. As stated in Section 3.3.1, the conventional use of the neutral point current i_n, as reconstructed by the phase current signals, (3.9), is problematic [26] since the measured currents are highly distorted at low switching frequency operation.

3.3.2.3 Performance of the redundant subbridge method

The performance of neutral point potential control is experimentally verified using the medium-voltage drive of the industrial implementation, Section 2.3. Its dc-link circuit is directly supplied through a 18-pulse diode rectifier from the 4.16-kV ac line without an intermediate transformer.

The algorithm for space vector modulation is programmed in a specific way to achieve minimum computational load. The method consists in going through a sequence of three logic decisions. The resulting logic state identifies the appropriate set of switching state vectors, and also instructs, by means of tabled rules, how the respective on-state durations are determined. The logic conditions are derived from the signs, or the amplitude relationship, of the two components of the voltage reference vector. The respective on-state durations are obtained by simply adding or substracting two magnitude values of the reference vector components, according to the respective tabled rule [27].

In the neutral point potential controller Fig. 3.16, the width of the tolerance band was set to ±40 V. This value amounts to double the neutral point potential ripple at steady-state as seen in Fig. 3.8, or to 1.4% of the dc-link voltage of 5.8 kV.

The dynamic response of the neutral point potential control scheme is shown in the oscillogram Fig. 3.17. The initial error of the neutral point potential is 400 V. The neutral point potential controller is engaged at t_1 = 10 ms. The error is then reduced to zero within 9 ms.

For comparison, Fig. 3.13 shows the performance of a state-of-the art neutral point potential scheme [21]. The manipulated variables are the on-durations of a selected pair of redundant switching state vectors. This method does not exploit the maximum controllability of the neutral point current and hence reacts slower. As a side effect, the volt-second commands per modulation subcycle are altered when a neutral point potential error exists. The consequences are increased harmonic distortions of the phase current waveforms. The effect was explained with reference to vector diagram Fig. 3.12.

The concept of redundant subbridge selection does not interfere with the modulation law, thus maintaining the same harmonic distortion of the phase currents that exists at steady-state. This is an important aspect for pulsewidth modulation at very low switching frequency.

4. Linear current controllers

Overview

In applications of limited dynamic performance, open-loop control is a preferred approach. Industrial fans or pumps are typical examples in the high-power drive range. The stator voltage of the induction motor is adjusted in proportion to the fundamental frequency, a method known as *v/f*-control. Electromechanical oscillations of the motor are avoided by operating with a gradient-limited reference input, and thus at quasi-fixed fundamental frequency f_1.

High-performance drives must respond to fast changes of velocity and torque. Such requirement is mostly met by imposing control on the current space vector i_s in a closed-loop. The general structure of closed-loop control is described in Section 4.1: the variables are referred to in field coordinates. The space vector of the fundamental stator current i_1 is employed as feedback signal.

Section 4.2 describes rotor field orientation, a technique that guarantees high control bandwidth by achieving decoupled control of the excitation and of the machine torque.

The system delay resulting from the digital processing system and the PWM inverter is described in Section 4.3. It is an important parameter for the design of closed-loop controllers in medium-voltage drives operating at low switching frequency.

Following the system modelling in the frequency domain, Section 4.4, a conventional PI current controller is designed in Section 4.5. An improvement of its behavior is attempted in Section 4.6 by employing feedforward control; the system performance is reevaluated. It is shown that feedforward control fails to fully compensate the undesired cross-coupling effect. A much better solution based on a controller having single complex poles is introduced in Section 4.7.

4.1 General structure of the closed-loop controller

The general principle of closed-loop control is illustrated in Fig. 4.1. The stator current space vector i_s is compared with a reference input i_s^* to form the stator current error

$$\Delta i_s = i_s^* - i_s. \tag{4.1}$$

The space vector Δi_s is the input to the current controller that generates the voltage reference vector u^* to drive the inverter. Fast elimination of

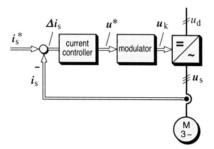

Fig. 4.1 Signal flow graph of an ac machine drive controlled in closed-loop; general structure

the error (4.1) is required in drives requiring high dynamic performance.

The closed-loop controller is commonly designed with the stator current space vector i_s referred to in the field coordinate system. The method is illustrated in the vector diagram Fig. 4.2. Expressions

$$i_d = \frac{2}{3}\left[cos\,\delta \cdot i_a + cos(\delta - 2\pi/3) \cdot i_b + cos(\delta + 2\pi/3) \cdot i_c\right] \qquad (4.2a)$$

$$i_q = -\frac{2}{3}\left[sin\,\delta \cdot i_a + sin(\delta - 2\pi/3) \cdot i_b + sin(\delta + 2\pi/3) \cdot i_c\right] \qquad (4.2b)$$

transform the three-phase currents i_a, i_b, and i_c to equivalent scalar current components i_d and i_q in dependency of the rotor field angle δ. The zero-sequence current component is equal to zero owing to the star connection arrangement of the machine windings [12].

As shown in the vector diagram Fig. 4.2, the scalar currents i_d and i_q in (4.2) are the respective projections of the stator current space vector current i_s on the real and imaginary axis of the field coordinate system (F). The stator current space vector describes the current density distribution inside the ac machine, whereas the external scalar quantities i_a, i_b, and i_c in (4.2) are measured at the machine terminals [12]. Scalar expressions (4.2) can be put in terms of space vector notation. The stator current space vector is denoted by $i_s^{(S)}$ when referred to in stationary coordinates; it is transformed to the field coordinate system by multiplying it with the unity vector rotator

$$i_s^{(F)} = exp(-j\delta) \cdot i_s^{(S)}, \qquad (4.3)$$

where $i_s^{(F)} = i_d + ji_q$.

44

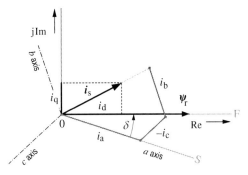

Fig. 4.2 Transformation of the three-phase current components i_a, i_b, and i_c to the two-phase scalar currents i_d and i_q

The signal flow graph Fig. 4.3 shows a current controller operating in field coordinates. It is observed in the figure that the space vector of the *fundamental* current i_1, rather than the total stator current i_s, is fed back to the input of the controller. Such selection is possible when the drive is operated at space vector modulation: the space vector i_1 is acquired by sampling the stator current at double the switching frequency f_s, as shown in Fig. 4.3. The method prevents the stator current harmonics from entering the control loop and from interfering with the control

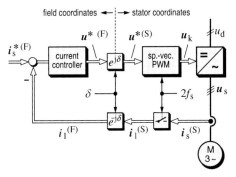

Fig. 4.3 Closed-loop control structure of a drive operated at space vector modulation. The field angle δ is employed for transformations between the stationary [(S)] and the field [(F)] coordinate system.

action: this is particularly important in medium-voltage drives operated at low switching frequency that produces relatively high current harmonics.

In the following, the symbol i_s of the total current space vector will be used to denote the feedback signal of the control loop, as is usually the case in design investigations. In a practical implementation, i_s is replaced by the respective fundamental component i_1.

4.2 Control with rotor field orientation

The design concept aims at a high control bandwidth by allowing the controller to displace the stator current space vector to any desired position in the complex plane: a fast current control system can override the stator winding dynamics. The dynamic order of the system represented in the signal flow graph Fig. 2.4 – shown on page 10 – then reduces. The system is only characterized by the single complex equation of the rotor winding

$$\tau_r \frac{d\psi_r}{d\tau} + \psi_r = -j\omega_r\tau_r\psi_r + l_m i_s. \qquad (4.4)$$

This technique is called control with *rotor field orientation*: as the real axis of the field coordinate system is aligned with the rotor flux space vector $\psi_r = \psi_{rd} + \psi_{rq}$, Fig. 4.2, the imaginary part of the rotor flux space vector is zero by definition, $\psi_{rq} = 0$. The single complex expression (4.4) then converts to the scalar equations

$$\tau_r \frac{d\psi_{rd}}{d\tau} + \psi_{rd} = l_m i_d \qquad (4.5a)$$

$$\omega_r \tau_r \psi_{rd} = l_m i_q. \qquad (4.5b)$$

Expression (4.5b) is referred to as the *condition for rotor field orientation*: it determines the value ω_r of the rotor (slip) frequency that is required to achieve decoupled control; the resulting scheme is illustrated in the signal flow graph Fig. 4.4. The scalar current components i_d and i_q act as forcing functions for the excitation and the machine torque, respectively.

4.3 System delay

The output of the current controller, the reference voltage u^*, is fed to the pulsewidth modulator to form the switching state vectors u_k that

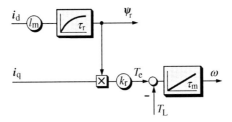

Fig. 4.4 Signal flow graph illustrating the decoupled torque-excitation control of an ac machine drive operating at rotor field orientation.

control the power inverter. The process does *not* have an immediate effect on the stator current space vector u_s owing to the system time delay τ_d that is invoked by the digital signal processing system and the PWM inverter. The system delay is visualized in the signal flow graph Fig. 4.5. In an inverter controlled by space vector modulation, the normalized delay $\tau_d = 1.5\ \omega_{sR}/(2f_s)$ [28], where $2f_s$ is the sampling frequency and f_s is the switching frequency.

The system delay τ_d determines the dynamic performance of a drive that is controlled in closed-loop [4]. In systems operating at high switching frequency f_s, a compensation method for the time delay is mostly omitted in the design of current controllers. Such simplification is not possible in medium-voltage drives operated at low switching frequency [29]: high system delays are generated that have a detrimental effect on the system performance.

The system delay τ_d is conventionally approximated by a first-order complex system

$$\tau_d \frac{du_s}{d\tau} + (1 + j\omega_s\tau_d)u_s = u^*. \tag{4.6}$$

The imaginary term $j\omega_s\tau_d$ in (4.6) results from the representation of the

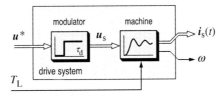

Fig. 4.5 Signal flow graph of the system delay

delay element in field coordinates.

4.4 System modelling in the frequency domain

The electromagnetic subsystem of the induction motor is described using the stator current vector i_s and the rotor flux linkage vector $\boldsymbol{\Psi}_r$ as state variables. A representation in field coordinates is chosen, Fig. 2.4, page 10. Transforming the continuous-time expressions (2.6), page 11 to the frequency domain yields the transfer function of the motor

$$F_m(s) = \frac{I_s(s)}{U_s(s)} \tag{4.7}$$

$$= \frac{\tau_r s + 1 + j\omega_r \tau_r}{r_\sigma(\tau_\sigma' s + 1 + j\omega_s \tau_\sigma')(\tau_r s + 1 + j\omega_r \tau_r) - k_1(1 - j\omega\tau_r)}$$

where $k_1 = k_r l_m/(r_\sigma \tau_r)$.

The system delay τ_d is approximated by

$$F_d(s) = \frac{U_s(s)}{U^*(s)} = \frac{1}{\tau_d s + 1 + j\omega_s \tau_d} \tag{4.8}$$

where $U_s(s)$ and $U^*(s)$ are the Laplace transforms of the respective output and input signals $u_s(\tau)$ and $u^*(\tau)$. The transfer function (4.8) is the Laplace transform of the continuous-time expression (4.6).

4.5 PI current control

The signal flow graph of the current control system in field coordinates is shown in Fig. 4.6. It is customary when designing the current controllers to neglect all signals that multiply by imaginary gain coefficients. The respective signal lines are shown in light gray in Fig. 4.6. Given this approximation, the respective imaginary and real components of the space vectors of the current loop result independent of each other. A PI controller is assigned to each current component. The respective transfer functions are

$$F_r(s) = \frac{U^*(s)}{\Delta I_s(s)} = g_c \frac{\tau_i s + 1}{\tau_i s}, \tag{4.9}$$

where τ_i is the integration time constant and g_c is the proportional controller gain. The open loop transfer function becomes

$$F_0(s) = \frac{g_c}{r_\sigma} \frac{\tau_i s + 1}{\tau_i s} \frac{1}{\tau_d s + 1} \frac{1}{\tau_\sigma' s + 1}. \tag{4.10}$$

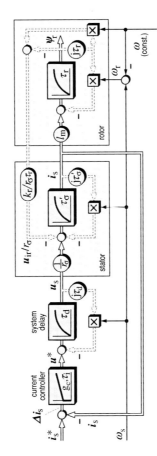

Fig. 4.6 Current controlled induction motor, signal flow graph. The signals represented by shaded lines are conventionally neglected in the controller design.

49

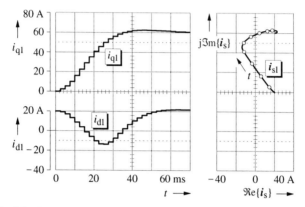

Fig. 4.7 Q-current step response, PI-controller; operation at rated speed and 200 Hz switching frequency, (a) fundamental current components, (b) trajectory of the fundamental current space vector, marked by circles at every 5 ms.

Letting $\tau_i = \tau_\sigma'$ eliminates the pole at $-1/\tau_\sigma'$. The gain g_c is adjusted for critical damping, $g_c = \tau_\sigma'/(2\tau_d)$, which yields the closed loop transfer function

$$F_c(s) = \frac{F_0(s)}{F_0(s)+1} = \frac{I_s(s)}{I_s^*(s)} = \frac{1}{2\tau_d^2 s^2 + 2\tau_d s + 1}. \qquad (4.11)$$

This design procedure, although based on an approximated transfer function, gives satisfactory results if the system delay is negligible, $\tau_d \ll \tau_\sigma'$. This applies for switching frequencies of several kHz. The system delay is then small, permitting the controller to immediately override the machine dynamics and to independently force the stator current space vector i_s to its commanded value.

The eigenbehavior of the machine cannot be suppressed if τ_d and τ_σ' have the same order of magnitude. This is demonstrated in the oscillogram Fig. 4.7, showing a step response of the q-current component at very low switching frequency, $f_s = 200$ Hz. Cross-coupling between the q- and d-current dominates the response. The peak value of the d-current exhibits a -170% deviation from its constant reference value. The effect leads to a temporary demagnetization of the machine. Such undesired behavior is the consequence of using an unsuited controller and neglecting the imaginary gain coefficients in its design.

50

4.6 PI current control with feedforward compensation

The dynamics of the current control loop can be improved by compensating the motion induced voltage $j\omega_s \tau_\sigma' i_s$ in the stator winding by a feedforward signal. Fig. 4.8 shows the signal flow graph. The signals represented by shaded lines are again neglected in the controller design. The feedforward signal is chosen as $j\omega_s \hat{l}_\sigma i_s$ which, including the gain factor $1/r_\sigma$ at the input of the stator delay element, amounts to $j\omega_s \tau_\sigma' i_s$, where $\hat{l}_\sigma$ is the estimated total leakage inductance. The feedforward term is selected to eliminate the influence of the motion induced voltage $-j\omega_s l_\sigma i_s$ The intention is to reduce cross-coupling. The problem was described in Section 2.1.4.

The system delay cannot be considered in this compensation. The delay is negligible if $\tau_d \ll \tau_\sigma'$ holds, which is only true at high switching frequency. The larger delay at low switching frequency leads to an incomplete compensation. It is seen in the signal flow graph Fig. 4.8 that the compensation signal gets delayed in the process of digital sampling before it is used for cross-coupling compensation. The compensation is therefore not fully effective. Parameter errors in $\hat{l}_\sigma$ may deteriorate the situation.

The performance of feedforward compensation at low switching frequency is illustrated in Fig. 4.9. Cross-coupling still exists with a temporary deviation of the d-current by to -125%. Conditions would turn worse if an error existed in the estimated leakage inductance $\hat{l}_\sigma$. Parameter sensitivity is a common demerit of feedforward control.

The deficiencies of this approach show even at relatively high switching frequency (e.g. $f_s = 1.1$ kHz, [4]). These shortcomings indicate that PI current control is not a viable solution when the inverter operates at low switching frequency.

4.7 Current controller with complex eigenvalues

The complex coefficients in the transfer functions of the induction motor (4.7) and the system delay element (4.8) suggest the design of a current controller having complex eigenvalues [4]. The procedure starts from the transfer function of the plant

$$F_p(s) = \frac{I_s(s)}{U^*(s)} = \frac{1}{\tau_d s + 1 + j\omega_s \tau_d}$$

$$\cdot \frac{\tau_r s + 1 + j\omega_r \tau_r}{r_\sigma (\tau_\sigma' s + 1 + j\omega_s \tau_\sigma')(\tau_r s + 1 + j\omega_r \tau_r) - k_1 (1 - j\omega \tau_r)} \qquad (4.12)$$

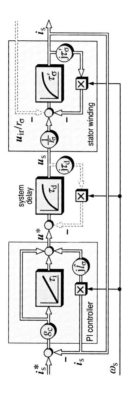

Fig. 4.8 Current control with feedforward compensation

52

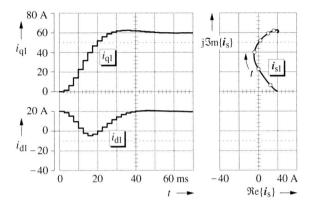

Fig. 4.9 Q-current step response, PI-controller with feedforward compensation; operation at rated speed and 200 Hz switching frequency, (a) fundamental current components, (b) trajectory of the fundamental current space vector, marked by circles at every 5 ms.

which includes the delay (4.8) and the machine (4.7). Based on this equation, a current controller is constructed as

$$F_r(s) = \frac{U^*(s)}{\Delta I_s(s)} = g_c r_\sigma \cdot \frac{\tau_d s + 1 + j\omega_s \tau_d}{(\tau_d s + 1)}$$

$$\cdot \frac{(\tau_\sigma' s + 1 + j\omega_s \tau_\sigma')(\tau_r s + 1 + j\omega_r \tau_r) - k_1(1 - j\omega\tau_r)}{\tau_i s(\tau_r s + 1 + j\omega_r \tau_r)} \qquad (4.13)$$

The controller is designed such that the poles and the zeroes of the plant are cancelled, and an integrating term is added for steady-state accuracy. The signal flow graph Fig. 4.10 illustrates that a compensating element is provided in the controller for every delay element of the plant, including the system delay.

The open-loop transfer function $F_o(s) = F_r(s) \cdot F_p(s)$, obtained from (4.12) and (4.13), results as a second-order system having real coefficients:

$$F_o(s) = F_r(s)F_p(s) = g_c \frac{1}{\tau_i s(\tau_d s + 1)}. \qquad (4.14)$$

This equation is identical to (4.10) and hence also (4.11) applies here. There is a difference, though, since no approximations have been made

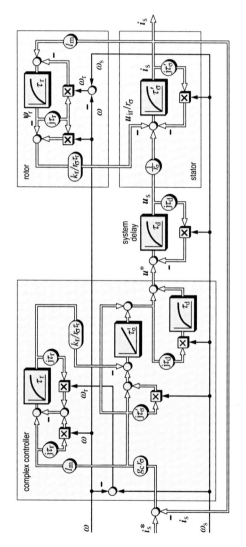

Fig. 4.10 Current control using a controller with complex gain coefficients

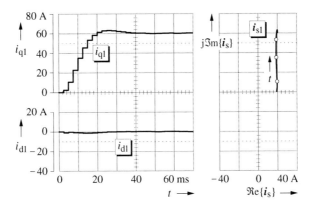

Fig. 4.11 Q-current step response, current controller with complex eigenvalues; operation at rated speed and 200 Hz switching frequency, (a) fundamental current components, (b) trajectory of the fundamental current space vector, marked by circles at every 5 ms.

in the controller design. Since the variables of the closed-loop transfer function (4.11) are of complex nature, the two eigenvalues of the characteristic equation (4.11) are *single complex* poles [4].

The transient response in Fig. 4.11 shows that cross-coupling is almost eliminated. All experimental results presented in this Chapter were conducted with the low-voltage prototype of the setup, Section 2.3.

5. Synchronous optimal modulation

Overview

Closed-loop control methods as commonly applied and presented in the preceding chapter assume operation of the power inverter at space vector modulation. The related dynamic problems with respect to response time and cross-coupling can be solved using a novel current controller having complex eigenvalues, as described in Section 4.7. Even when operated at very low switching frequency, the system can achieve high dynamic performance.

Setting the switching frequency to values well below 1 kHz is a priority for medium-voltage drives: it allows minimizing the switching losses of the semiconductors of the power inverter. In Section 5.1 it is shown that such objective cannot be satisfied by space vector modulation: the reduction of the switching frequency is accompanied by unacceptable high harmonic currents in the machine windings.

For this reason, the selection of the pulsewidth modulation method is reevaluated in the present Chapter. Very low switching frequency can be only maintained by synchronous optimal modulation. In this method, the pulse patterns are calculated in an off-line procedure such that the switching frequency is *synchronized* to the fundamental period of the voltage waveform. The respective switching angles are *optimized* over one fundamental period at each steady-state operating point by observing the harmonic current distortion; the algorithm is detailed in Section 5.2.

The modulator presented in Section 5.3 converts the optimized switching angles to respective time instants in a real-time implementation. It allows operating the system from a given reference input, steady-state conditions assumed.

Section 5.4 presents experimental results recorded from the medium-voltage drive of the industrial setup, which was described in Section 2.3. A comparison between synchronous optimal modulation and space vector modulation is carried out to validate the suitability of synchronous optimal modulation. Achieving operation at very low switching frequency with the harmonic currents kept at minimum values allows maximizing the fundamental output power of the PWM inverter; this fact is explained in Section 5.5.

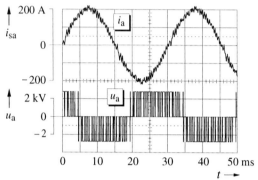

Fig. 5.1 Space vector modulation at 1 kHz switching frequency, f_1 = 33.3 Hz, m = 0.67; measured waveforms of the current and the output potential, both of phase a.

5.1 Current harmonics

The oscillograms in Fig. 5.1 were recorded from the medium-voltage drive by assuming operation at space vector modulation. The fundamental frequency f_1 = 33.3 Hz and modulation index m = 0.67. The three-level output voltage potential of phase a is shown in the lower trace of Fig. 5.1. The upper trace of Fig. 5.1 is the respective current waveform i_a. In these oscillograms, the switching frequency f_s = 1 kHz which procudes currents of low harmonic content.

Conditions change if the switching frequency is reduced in order to restrain the switching losses of the power semiconductor devices. In the oscillographed steady-state waveforms of Fig. 5.2, the operating point of the medium-voltage drive is set again to f_1 = 33.3 Hz and m = 0.67, while the switching frequency is f_s = 200 Hz, reduced to one-fifth compared to Fig. 5.1. The current waveform in the upper trace of Fig. 5.2 appears highly distorted. Such performance is unacceptable, because it leads to high losses in the machine windings.

Even though space vector pulsewidth modulation produces the lowest harmonic distortion among carrier modulation schemes, [30], it is unsuited for medium-voltage applications. Operation at reduced switching frequency requires appropriate modulation techniques in order to contain the steady-state current harmonics.

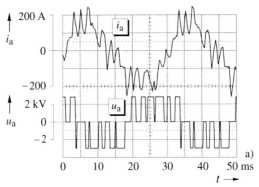

Fig. 5.2   Space vector modulation at 200 Hz switching frequency, $f_1 = 33.3$ Hz, $m = 0.67$; measured waveforms of the current and the output potential, both of phase a.

5.2   Optimized pulse patterns

5.2.1   Objective function

An optimization of the pulsewidth-modulated pulse patterns is possible under the restriction that steady-state conditions exist. The approach implies that switching frequency and fundamental frequency are synchronized. Their ratio, the pulse number $N = f_s / f_1$, is then an integer number. Satisfying an objective function, a set of switching angles per fundamental period T_1 is determined for every steady-state operating point. A preferred objective is the minimization of the total harmonic distortion of the machine currents [6],

$$i_{\mathrm{h\,rms}} = \sqrt{\frac{1}{T_1} \int_{T_1} i_{\mathrm{h}}^2(\tau) d\tau} \to min, \qquad (5.1)$$

where i_{h} is the space vector of the harmonic current generated at operating point (m, N). The respective switching patterns are precalculated and stored in a memory of the controlling microprocessor as functions of the modulation index m and the pulse number N.

5.2.2   Optimization process

The analysis concerns a three-phase induction machine fed by a three-

59

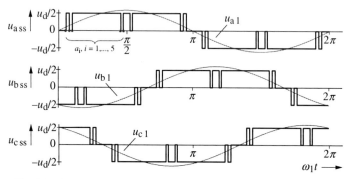

Fig. 5.3 Three-phase, steady-state, switched phase potentials $u_{a\,ss}$, $u_{b\,ss}$, and $u_{c\,ss}$ in a three-level inverter drive system plotted over the phase angle $\omega_1 t$. The waveforms are generated by synchronous optimal modulation. The fundamental frequency $f_1 = 40$ Hz, the switching frequency $f_s = 200$ Hz, and the pulse number $N = 5$. The upper trace of the figure shows the commutation angles a_i, $i \in \{1, 2, ..., 5\}$ in the first quarter of the phase-a waveform $u_{a\,ss}$.

level inverter. The waveforms of the three-phase steady-state phase potentials exhibit periodicity $T_1 = 2\pi/\omega_1$, where $\omega_1 = 2\pi f_1$. Their instantaneous values $u_{a\,ss}$, $u_{b\,ss}$, $u_{c\,ss} \in \{-u_d/2, 0, +u_d/2\}$, where the subscript $_{ss}$ refers to steady-state. The potential of phase b voltage lags phase a by $2\pi/3$,

$$u_{b\,ss}(\omega_1 t) = u_{a\,ss}(\omega_1 t - 2\pi/3), \qquad (5.2a)$$

while the phase c potential leads phase a by $2\pi/3$,

$$u_{c\,ss}(\omega_1 t) = u_{a\,ss}(\omega_1 t + 2\pi/3). \qquad (5.2b)$$

The potential waveform of phase a exhibits quarter-wave symmetry

$$u_{a\,ss}(\omega_1 t) = u_{a\,ss}(\pi - \omega_1 t), \qquad 0 \le \omega_1 t \le \pi/2 \qquad (5.3a)$$

$$u_{a\,ss}(\omega_1 t) = -u_{a\,ss}(\omega_1 t - \pi), \qquad 0 \le \omega_1 t \le \pi. \qquad (5.3b)$$

The steady-state potential waveforms $u_{a\,ss}$, $u_{b\,ss}$, and $u_{c\,ss}$ of one fundamental period T_1 are illustrated in Fig. 5.3. The pulse number $N = f_s/f_1$ equals 5 in this example. The location of the switching angles a_i, $i \in \{1, 2,..., 5\}$ over the first quarter of the voltage waveform $u_{a\,ss}$ are indicated in the upper trace of the figure.

The optimization algorithm sets off by evaluating the Fourier Series

60

of the switched phase potential $u_{a\,ss}$, expressed in the form

$$u_{a\,ss}(\omega_1 t) = \sum_{k=0}^{+\infty} u_{a\,ss}(k)\,sin(k\omega_1 t). \tag{5.4}$$

Owing to the synchronism between the pulse pattern and the fundamental waveform, *subharmonic* spectral components do not exist. As quarter-wave symmetry (5.3) is assumed, all harmonics of even order and all triplen harmonics are zero. With the remaining harmonics of integer order $k = 1$ and $K = \{5, 7, 11, 13, 17, 19, ...\}$, expression (5.4) converts to

$$u_{a\,ss}(\omega_1 t) = \sum_{k\in\{1,K\}} u_{a\,ss}(k)\,sin(k\omega_1 t), \tag{5.5}$$

where $k = 1$ corresponds to the fundamental component. The variable $u_{a\,ss}(k)$ denotes the magnitude of the k-th harmonic of the a-phase potential waveform. It is evaluated by

$$u_{a\,ss}(k) = \frac{4}{T_1} \int_0^{T_1} u_{a\,ss}(t)\,sin(k\omega_1 t)dt \tag{5.6}$$

to yield

$$u_{a\,ss}(k) = \frac{4u_d}{\pi} \sum_{k\in\{1,K\}} \frac{1}{k}\left(\sum_{i=1}^{N}(-1)^{i+1} cos(ka_i)\right). \tag{5.7}$$

The steady-state *harmonic* content of the a-phase potential waveform results from the components of subset $k \in K$ in (5.7),

$$u_{ah\,ss}(k) = \frac{4u_d}{\pi} \sum_{k\in K} \frac{1}{k}\left(\sum_{i=1}^{N}(-1)^{i+1} cos(ka_i)\right). \tag{5.8a}$$

The *fundamental* component is retrieved by assigning $k = 1$ in (5.7),

$$u_{a1} = \frac{4u_d}{\pi} \sum_{i=1}^{N}(-1)^{i+1} cos(a_i). \tag{5.8b}$$

Expression (5.8b) is then normalized by the fundamental component of the six-step mode. Observing the potential waveforms shown in Fig. 2.10, page 21, the pulse number $N = 1$ and the single commutation angle $a_1 = 0$ at six-step mode. The expression (5.8b) therefore yields

$$u_{a1\,six\text{-}step} = \frac{4u_d}{\pi}. \tag{5.9}$$

Referring to (5.8b), (5.9), and to the definition (2.17), page 22, the modulation index is expressed as function of the switching angles a_i

$$m = \sum_{i=1}^{N} (-1)^{i+1} \cos(a_i).$$ (5.10)

If one degree of freedom is reserved to satisfy (5.10), the remaining $N-1$ degrees of freedom are left for the optimization procedure.

To satisfy the objective function (5.1), the switching angles a_i are selected so as to minimize the rms harmonic current of the phase current

$$i_{\text{ah rms}} = \sqrt{\sum_{k \in K} i_{\text{ah ss}}^2(k)}.$$ (5.11)

The discrete-frequency, steady-state harmonic current $i_{\text{ah ss}}(k)$ in (5.11) is the ratio of the harmonic content of the steady-state phase potential $u_{\text{ah ss}}(k)$ over the magnitude of the respective complex impedance $z(k)$ $= k\omega_1 l_\sigma$ [31],

$$i_{\text{ah ss}}(k) = \frac{u_{\text{ah ss}}(k)}{z(k)}.$$ (5.12)

Inserting expression (5.12) in (5.11) yields

$$i_{\text{ah rms}} = \frac{1}{l_\sigma} \sqrt{\sum_{k \in K} \left(\frac{u_{\text{ah ss}}(k)}{k\omega_1} \right)^2}.$$ (5.13)

The rms harmonic current (5.13) is normalized by its value at six-step mode to eliminate the term of the machine-dependent leakage inductance l_σ. The *distortion factor* is thus defined

$$d = \frac{i_{\text{ah rms}}}{i_{\text{ah rms six-step}}}.$$ (5.14)

The switching angles a_i are sought iteratively in the interval $\omega_1 t \in (0, \pi/2)$ such that the distortion factor d is minimum for each discrete value of the modulation index m [6]. A practical solution to such problem is a numerical implementation of the gradient descent algorithm [32]. The respective algorithm was developed on a 2.5 GHz single-core microprocessor: results were obtained for the optimized pattern at pulse number $N = 2$ within 2 min. Calculation times increase drastically at high pulse numbers: more than 180 min were required to optimize the pulse pattern at $N = 12$.

5.2.3 Result of the optimization

The curves in Fig. 5.4 represent the optimal switching angles versus the modulation index m for the four lowest values of the pulse number N.

The set of optimal pulse patterns for a particular implementation is obtained by selecting a maximum value $f_{s\,max}$ of the switching frequency and by observing the condition $m = f_1/f_{1R}$ for constant flux operation in the base speed range. These conditions define the respective range of the modulation index m in which a pulse pattern N is installed. The result of such composition is shown in Fig. 5.5 for $f_{s\,max} = 200$ Hz and $f_{1R} = 50$ Hz. The value of $f_{s\,max}$ was selected as a trade-off between switching losses, harmonic distortion and utilization of the power semiconductor devices. Synchronous optimal modulation is applied in the range above $m = 0.3$; below this value, space vector modulation is employed. The reason for such selection is explained in Section 5.4.

The distortion factor d that corresponds to the optimized patterns is mapped in Fig. 5.5(c) in a logarithmic scale. It assumes its maximum value $d = 1$ in the six-step mode at $m = 1$. At any other discrete value of the modulation index, the minimum distortion factor is produced as the result of the optimization.

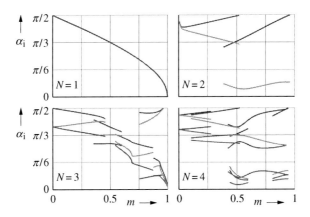

Fig. 5.4   Optimal switching angles a_i at different pulse numbers N

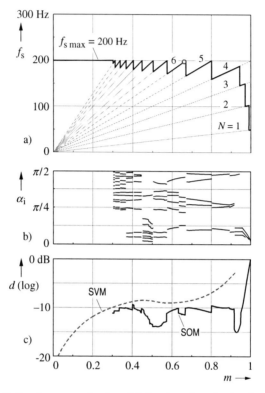

Fig. 5.5 Synchronous optimal modulation in a three-level voltage source inverter of rated fundamental frequency $f_{1R} = 50$ Hz; (a) switching frequency f_s and pulse number N over the modulation index m, (b) optimized switching angles in the range $m \in [0.3, 1]$, and (c) harmonic distortion factor d over m. The dashed line shows the distortion factor under space vector modulation for comparison. The maximum value of the switching frequency $f_{s\,max} = 200$ Hz.

5.3   The synchronous optimal pulsewidth modulator

The optimization process is carried out off-line. The switching angles α_i, $i \in 1 \ldots N$, are precalculated under assumed steady-state conditions. They are stored in a memory table as complete pulse patterns P.

While the drive is in operation, the magnitude u^* of the reference voltage vector u^* is used to select the appropriate optimal switching pattern. The patterns form closed sets of switching angles which are referred to as $P(m, N)$ in the signal flow graph Fig. 5.6. The notation $P(m, N)$ indicates that they are functions of the modulation index m and the pulse number N. The modulation index is proportional to u^*. The phase angle $arg(u^*)$ of the voltage reference vector identifies the required phase displacement within the selected pattern $P(m, N)$, while the signal f_1 of the fundamental frequency converts the respective switching angles to time instants, $t_i = a_i/(2\pi f_1)$. Receiving these data, the modulator Fig. 5.6 reconstructs the optimal sequence of discrete switching state vectors u_k in real-time.

In Fig. 5.7, the three-phase phase potentials are shown; the pulse number $N = 5$ in the first quarter of one fundamental period. The modulation index $m = 0.8$ and the fundamental frequency $f_1 = 40$ Hz. The switching frequency $f_s = f_1 \cdot N = 200$ Hz. The switching waveforms of the three phase potentials are shown in Fig. 5.7(a). As an example, the commutation angle a_1 translates to the switching instant $t_1 = a_1/(2\pi f_1)$.

The commanded switching state vectors u_k are shown in the lower portion of Fig. 5.7(a). The corresponding transitions in the complex plane as shown in Fig. 5.7(b).

5.4   Comparison to space vector modulation

The oscillographed waveforms in Fig. 5.8 are obtained with operation at synchronous optimum modulation. The fundamental frequency $f_1 = 33.3$ Hz, and the modulation index $m = 0.67$; the pulse number $N = 6$ and the switching frequency $f_s = 200$ Hz. The respective operat-

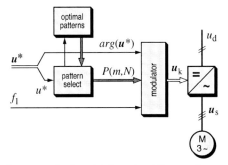

Fig. 5.6   Synchronous optimal modulation, signal flow graph

65

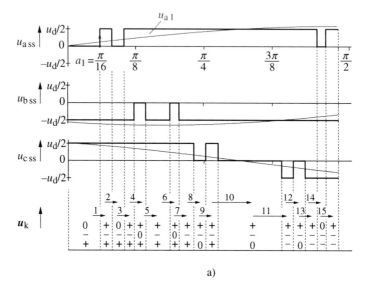

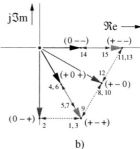

Fig. 5.7 Optimal pulse pattern $P(m = 0.8,$ $N = 5)$ at fundamental frequency $f_1 = 40$ Hz and switching frequency $f_s = 200$ Hz; (a) three-phase potentials and switching state vectors u_k vs. one quarter of the fundamental period, (b) respective transitions between the selected switching state vectors u_k shown in the complex plane.

ing point is marked by a small circle in the upper trace of Fig. 5.5(a). The current waveform in the upper trace of Fig. 5.8 is compared to the respective waveform in Fig. 5.2, recorded in the same operating point using space vector modulation. The comparison shows that synchronous optimal modulation produces much lower harmonic content of the stator current.

This result is confirmed by the current harmonic trajectories in Fig. 5.9, obtained at modulation index $m = 0.85$. The current harmonic tra-

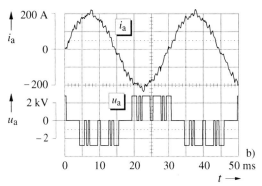

Fig. 5.8 Measured waveforms of the stator current i_a, and of the inverter output potential u_a, switching frequency 200 Hz, fundamental frequency 33.3 Hz; modulation index $m = 0.67$; operation at synchronous optimal modulation

jectory at space vector modulation (SVM) is shown over one subcycle T_0; the magnitude of the harmonic current $i_{h\,ss}$ amounts to 40% of the rated machine current. In comparison, the trajectory at synchronous optimal modulation (SOM), extending over one sixth of the fundamental period, has a much lower harmonic content.

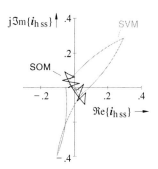

Fig. 5.9 Trajectories of the steady-state harmonic current $i_{h\,ss}$ using space vector modulation (SVM) and synchronous optimal modulation (SOM), respectively; modulation index $m = 0.85$.

Further conclusions may be drawn with respect to Fig. 5.5. It is shown there that synchronous modulation is applied in the modulation index range $m > 0.3$ including *overmodulation* [33]. This is an inherent property of the modulation technique. Contrary to this, space vector modulation achieves a maximum modulation index value of only $m = 0.907$ [34]; operation in the overmodulation range necessitates modified space vector modulation techniques [35].

The solid line in Fig. 5.5(c) shows the distortion factor d

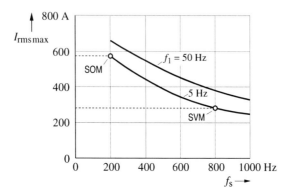

Fig. 5.10 Maximum rms current vs. switching frequency with the fundamental frequency f_1 as a parameter; $EUPEC$ 6.5-kV, 600-A IGBT; possible operating points at carrier modulation and optimal modulation are marked by circles.

plotted at synchronous optimal modulation (SOM) versus the modulation index at $m \in [0.3, 1]$. The dashed line is the distortion factor at space vector modulation (SVM); it covers the low modulation index range, including values below $m = 0.3$. As confirmed from the plot, it produces low distortion factor when $m < 0.3$ [5]. Using space vector modulation is therefore preferred in this range. Such selection is also legalized by the complexity of implementation of synchronous optimal modulation at high pulse numbers.

In the modulation index range $m \in [0.3, 1]$ where synchronous optimal modulation is applicable, its superiority over space vector modulation is confirmed by the plotted curves of the distortion factor in Fig. 5.5(c).

5.5   Reduction of the switching losses

The reduced harmonic content of synchronous optimal modulation permits operation at very low switching frequency. The switching losses then reduce, and the maximum load current can be increased. The trade-off between switching losses and conduction losses of the power semiconductors is interpreted in the graph Fig. 5.10. It shows the results of a device loss simulation [36] for the $EUPEC$ 6.5-kV, 600-A IGBT. Operation at 3.6 kV dc link voltage and 80° chip temperature was assumed. A viable design concept reduces the switching frequency from 800 Hz at carrier modulation to 200 Hz at synchronous optimal modulation. Such reduction of the switching frequency conserves the harmonic distortion, while reducing the switching losses to only 25%. This permits increasing the maximum load current. The fundamental output power of a three-level inverter based on 6.5-kV IGBTs results more than doubled.

6. Dynamic modulation error

Overview

Optimal pulse patterns are determined by off-line calculation, assuming steady-state operation of the machine drive, Section 5.2 [7, 9]. Fed by a voltage reference vector u^* of constant magnitude, the modulator Fig. 5.6, page 65, perpetuates a fixed operating point (m, N) by selecting periodically the respective optimized pulse pattern $P(m,N)$. As described in Section 6.1, the resulting stator current develops on a trajectory of optimal steady-state harmonic content $i_{h\,ss}$.

Operation of the drive system in quasi steady-state entails variations of the voltage reference vector u^*. Transitions between neighboring pulse patterns are invoked as a consequence. They interfere adversely with the optimal modulation patterns: the stator current harmonics deviate from their preoptimized trajectory $i_{h\,ss}$. A dynamic modulation error is defined in Section 6.2 to measure this deviation. As a consequence of the dynamic modulation error, high overcurrents are encountered [8].

6.1 The steady-state current trajectory

The voltage reference vector u^* follows a circular trajectory when the operating point (m, N) is fixed. The synchronous optimal pulsewidth modulator Fig. 5.6, page 65, selects the respective pulse pattern $P(m,N)$. The pattern comprises all preoptimized steady-state switching instants that occur during a fundamental period. The instants t_i, where $i = 1 \ldots 12 \cdot N$ in a three-level inverter, and the respective instantaneous values of the phase potentials define three-phase, steady-state potential waveforms $u_{a\,ss}$, $u_{b\,ss}$, and $u_{c\,ss}$, an example of which was given in Fig. 5.3, page 60. In space vector representation

$$u_{ss} = \frac{2}{3}(1 \cdot u_{a\,ss} + a u_{b\,ss} + a^2 u_{c\,ss}) \tag{6.1}$$

where $u_{a\,ss}$, $u_{b\,ss}$, $u_{c\,ss} \in \{-u_d/2, 0, +u_d/2\}$.

Fed to the induction machine through the PWM inverter, the steady-state voltage u_{ss} generates a stator current waveform of optimal harmonic content. Observed over the duration of one fundamental period, the stator current space vector

$$i_{ss} = i_1 + i_{h\,ss} \tag{6.2}$$

develops on an optimum *steady-state trajectory*. The space vector i_1 in

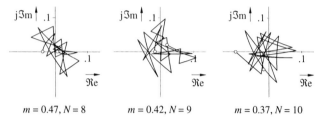

$j\Im m \uparrow$ .1	$j\Im m \uparrow$ .1	$j\Im m \uparrow$ .1
.1	.1	.1
$\Re e$	$\Re e$	$\Re e$
$m = 0.47, N = 8$	$m = 0.42, N = 9$	$m = 0.37, N = 10$

Fig. 6.1 Synchronous optimal modulation; trajectories of the steady-state harmonic current $i_{h\,ss}$ at different values of modulation index m and pulse number N. Only one sixth of a fundamental period is displayed.

(6.2) is the fundamental stator current and $i_{h\,ss}$ is the space vector of the steady-state harmonic current.

Having assumed steady-state conditions, the load torque of the electromechanical system is constant. Deviations from steady-state of the magnitude and phase angle of the fundamental current vector i_1 are therefore not considered by the optimization: the pulse pattern $P(m, N)$ optimizes exclusively the harmonic content of the stator current (5.14), page 62.

The trajectory of the steady-state harmonic current $i_{h\,ss}$, and the respective steady-state current trajectory i_{ss}, have inherited two properties from the optimal pattern:

• They refer to steady-state operation, and
• they are optimal trajectories.

Steady-state harmonic trajectories $i_{h\,ss}$ in different operating points of synchronous optimal modulation are shown in Fig. 6.1 with the modulation index m and the pulse number N as parameters. Only one sixth of a fundamental period is displayed. The respective initial and final values are marked by circles in Fig. 6.1. They locate in phase displacements of 60° with respect to the origin. Owing to its synchronization with the fundamental, the entire harmonic trajectory follows a closed pattern. It is formed by adding the other five 60°-sections, each rotated by 60° with respect to the previous section. As an example, a complete harmonic trajectory is shown in Fig. 6.2(b) for $m = 0.94$ and $N = 3$; one 60°-section is given in Fig.6.2(a). The respective total steady-state current trajectory i_{ss} is shown in Fig. 6.2(c). The dotted line in Fig. 6.2(c) is the fundamental current trajectory i_1.

Fig. 6.3 is an oscillogram recorded from the 2.5-MVA inverter drive

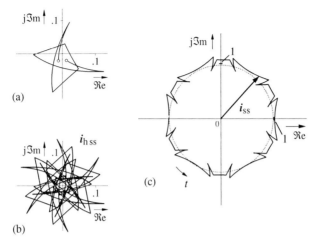

Fig. 6.2 Steady-state current trajectories at $N = 3$, synchronous optimal modulation; (a) 60°-section of the harmonic trajectory, (a) complete optimal harmonic trajectory $i_{h\,ss}$, (b) steady-state current trajectory i_{ss}; the dotted line is the fundamental current component i_1.

of the industrial imlementation. The graph shows the trajectories of the stator current vector at rated load and at no-load, both at steady-state operation.

6.2 Harmonic deviations at quasi steady-state operation

In the following, it is assumed that the drive is operated at *quasi* steady-state: variations of the operating point may invoke small changes of the modulation index and of the fundamental frequency, while the fundamental current of the machine is almost constant. Following a commanded variation of the modulation index Δm, e.g. at time instant t_c, the reference voltage vector $\boldsymbol{u}^*$ is displaced to a neighboring location in space. A different pulse pattern is then instantaneously selected for inverter control in which the switching transitions appear at different time instances. The complex volt-second input to the machine that follows does no longer satisfy the optimum condition of minimum harmonic distortion.

This effect shall be further explained. It is assumed that the modulator operates at steady-state, being controlled by a particular pulse pat-

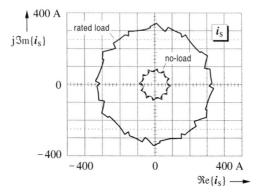

Fig. 6.3 Steady-state trajectories of the stator current vector; 2.5-MVA inverter drive, $f_1 = 42.5$ Hz, $f_s = 170$ Hz, $m = 0.86$, $N = 4$

tern $P_1(m_1, N_1)$. The stator current develops on a periodic steady-state trajectory T_{ss1} that is optimal and exhibits minimum harmonic distortion.

It is now assumed that a change of operating conditions occurs at time instant t_c. The pulse pattern instantaneously changes to $P_2(m_2, N_2)$, which, at steady-state, would produce the periodic stator current trajectory T_{ss2}.

Since the steady-state trajectories T_{ss1} and T_{ss2} have different harmonic content, the steady-state currents $i_{ss}(t_c^-) = i_{ss1}(t_c^-)$ and $i_{ss}(t_c^+)$ $= i_{ss2}(t_c^+)$ must differ. The *actual* stator current, however, keeps its value, $i_s(t_c^+) = i_s(t_c^-) = i_{ss}(t_c^-)$. Hence a dynamic modulation error

$$d_i(t_c) = i_{ss}(t_c^+) - i_s(t_c^-), \qquad (6.3)$$

occurs at t_c; the subscript $_i$ characterizes d_i as a *current* variable. The error (6.3) is defined as the deviation of the stator current space vector $i_s(t_c^-)$ at the instant of a commanded change from its optimal trajectory $i_{ss}(t_c^+)$ that becomes effective after the change under steady-state condition. Since $i_s(t_c^-) = i_{ss}(t_c^-)$, the error (6.3) may be equivalently expressed by the difference of the respective steady-state current trajectories

$$d_i(t_c) = i_{ss}(t_c^+) - i_{ss}(t_c^-). \qquad (6.4a)$$

Since the fundamental current is constant, the definition (6.4a) is con-

74

verted by (6.2) to the difference of the steady-state *harmonic* trajectories that are valid before and after the change of the operating point

$$d_i(t_c) = i_{h\,ss}(t_c^+) - i_{h\,ss}(t_c^-).$$ (6.4b)

The dynamic modulation error appears instantaneously at t_c. It adds to the steady-state harmonic current of the new pulse pattern $P_2(m_2, N_2)$.

Expression (6.4b) is exemplified with reference to Fig. 6.4. Operation at pulse number $N = 4$ is assumed in the waveforms of Fig. 6.4(a) with the modulation index $m = 0.93$. The a-phase steady-state potential waveform $u_{a\,ss}^{(I)}$ is displayed in the upper trace of the figure: it results from the periodic selection of the respective optimized pattern $P(0.93, 4)$. In the same trace, the respective fundamental voltage waveform $u_{a1}^{(I)}$ is shown. The difference $u_{ah\,ss}^{(I)} = u_{a\,ss}^{(I)} - u_{a1}^{(I)}$, is the steady-state harmonic voltage waveform of phase a; it is shown in the middle trace of Fig. 6.4(a). The lower trace of the figure is the waveform of the steady-state harmonic current $i_{ah\,ss}^{(I)}$.

The operating point changes at time instant $t_c = \pi/\omega_1$; a small increase of the modulation index by $\Delta m = 0.01$ forces the synchronous optimal modulator to select the new pulse pattern $P(0.94, 3)$ of lower pulse number, $N = 3$. As a result of the commanded transition, a new steady-state potential waveform is applied to phase a; it is displayed in the upper trace of Fig. 6.4(b) and it is denoted by $u_{a\,ss}^{(II)}$. Having assumed quasi steady-state operation, the new fundamental waveform $u_{a1}^{(II)} \cong u_{a1}^{(I)}$ in good approximation.

The newly selected pulse pattern $P(0.94, 3)$ has its optimized switching angles at different locations compared to the original pattern $P(0.93, 4)$. The waveform of the steady-state harmonic current $i_{ah\,ss}^{(II)}$ is also different as a consequence. It is shown in the lower trace of Fig. 6.4(b). The difference of the steady-state harmonic waveforms

$$d_{a\,i}(t_c) = i_{ah\,ss}^{(II)}(t_c^+) - i_{ah\,ss}^{(I)}(t_c^-)$$ (6.5)

is the a-phase component of the dynamic modulation error d_i that the commanded condition generates; it is shown in Fig. 6.4(c).

A change of operating conditions generates excursions of the harmonic current in all three phases of the system. The scalar components $d_{a\,i}$, $d_{b\,i}$, and $d_{c\,i}$ are the projections of the space vector

$$d_i = \frac{2}{3}(1 \cdot d_{a\,i} + a\,d_{b\,i} + a^2 d_{c\,i}).$$ (6.6)

on the respective phase directions. Fig. 6.5 shows the space vector of

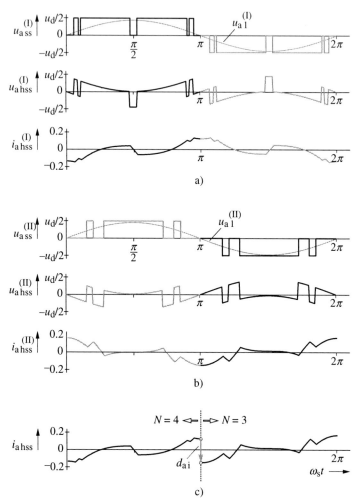

Fig. 6.4 Generation of dynamic modulation error in the direction of phase a, (a) steady-state voltage waveform $u_{a\,ss}^{(I)}$, harmonic voltage $u_{ah\,ss}^{(I)}$ and harmonic current $i_{ah\,ss}^{(I)}$ at pulse number $N = 4$, (b) same as in (a) with the pulse number set to $N = 3$, (c) dynamic modulation error d_{ai} generated by the transition from $N = 4$ to $N = 3$ at time instant $t_c = \pi/\omega$.

the dynamic modulation error d_i that is generated by the process in Fig. 6.4.

The effect of the dynamic modulation error was recorded in the low-voltage prototype drive of the experimental setup. First, the drive was subjected to steady-state operation with no load applied to the machine. The current tracked the optimized steady-state trajectory at pulse number $N = 6$, Fig. 6.6. The fundamental current magnitude $I_1 = 30$ A, the fundamental frequency $f_1 = 30.5$ Hz and the switching frequency $f_s = 183$ Hz.

Following a variation of the operating point at time instant $t = t_c$, the modulation index m changes. Even though the fundamental current stays unaffected, the harmonic content of the current deviates instantaneously from the optimized steady-state trajectory; the effect is shown in Fig. 6.7(a). The magnitude of the resulting dynamic modulation error at the time instant $t = t_c$ amounts to $d_i = 39$ A. The dynamic modulation error establishes a dc offset which decays in time as commanded by the transient machine time constant $\tau_\sigma' = l_\sigma/r_\sigma$ [8]. The process is illustrated in Fig. 6.7(b).

More errors may occur if the reference voltage keeps changing. New error vectors then add to already existing errors from previous changes. Particularly at low switching frequency, where the pulse durations are large, a deviation from the steady-state trajectory can cause high current transients. The magnitude of an accumulated total error may eventually increase to values as high as to provoke an overcurrent trip [7, 9].

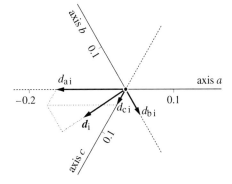

Fig. 6.5 Space vector of the dynamic modulation error d_i in the complex plane; transient process as described in Fig. 6.4.

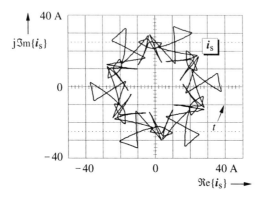

Fig. 6.6 Steady-state trajectory of the stator current vector in stationary coordinates, $f_1 = 30.5$ Hz, $m = 0.625$ and $N = 6$.

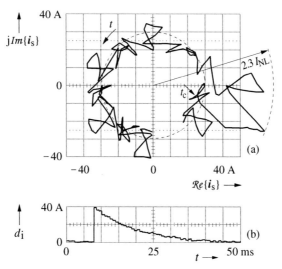

Fig. 6.7 Effect of the dynamic modulation error d_i at transient condition;
(a) stator current trajectory, (b) magnitude of the dynamic modulation error.
I_{NL}: no-load current.

78

7. Stator flux trajectory tracking

Overview

Even at quasi steady-state, the optimization of the pulse patterns produced by synchronous optimal modulation does not always hold. Control of the harmonic current trajectory is necessary in order to allow quasi steady-state operation. In Section 7.1, the harmonic model of the induction machine is explained: it associates the harmonic content of the stator voltage to the respective harmonic current. The approach allows real-time measurement of the dynamic modulation error – a first step in the process to establish control of the harmonic excursions.

Section 7.2 shows that the definition of the dynamic modulation error is preferably based on the harmonics of the stator *flux linkage* vector to avoid dependency from machine parameters. A practical evaluation of the error is given in Section 7.3.

Trajectory tracking control is employed in Section 7.4 to eliminate the dynamic modulation error. An optimal trajectory of the stator flux linkage vector is derived from the pulse pattern in actual use. The stator flux linkage vector is forced to follow this target trajectory. Modifying the stator flux trajectory enables closed-loop control of the stator flux harmonics in a deadbeat fashion while conserving optimal modulation.

7.1 The harmonic machine model

The dynamic modulation error was defined in (6.3), page 74, as

$$d_i = i_{ss} - i_s,$$ (7.1)

where i_{ss} is the optimal steady-state current trajectory produced by the pulse pattern in use and i_s is the actual stator current trajectory. Having assumed quasi steady-state operation, the fundamental current is considered constant. Definition (7.1) can be therefore expressed by the difference in the harmonic currents

$$d_i = i_{hss} - i_h.$$ (7.2)

Expression (7.2) implies that a method to evaluate the harmonic content of the current is required. The simplified model of the induction machine Fig. 7.1 is employed for this purpose. In Fig. 7.1(a), the machine is fed by the space vector of the total stator voltage u_s to generate stator current i_s. The machine parameters are the total leakage inductance l_σ and the equivalent resistance r_σ.

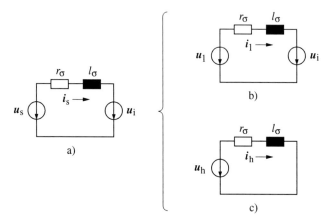

Fig. 7.1 Simplified induction machine model; (a) equivalent circuit,
(b) fundamental component, (c) harmonic component.

Expression

$$u_s = u_1 + u_h \tag{7.3}$$

decomposes the stator voltage to its fundamental and harmonic component. The respective equivalent circuits of Fig. 7.1(b) and 7.1(c) result.

Assuming constant stator frequency, the induced voltage only depends
on the fundamental of the stator flux linkage

$$u_i = j\omega_s \psi_1 . \tag{7.4}$$

Being a fundamental quantity, u_i is only included in the circuit Fig. 7.1(b).

The equivalent circuit Fig. 7.1(c) is valid for the harmonic response
of the induction machine: an excitation described by the harmonic voltage u_h generates the respective harmonic current i_h. Their relationship
is of first-order; it is characterized by the transient time constant $\tau_\sigma' = l_\sigma / r_\sigma$. At excitation frequencies higher than the corner frequency $1/\tau_\sigma'$,
the behavior of the harmonic model is this of an integrator of the har-
monic voltage

$$i_h = \frac{1}{l_\sigma} \int u_h \, dt . \tag{7.5}$$

The same process can be followed to associate the space vector of the

80

steady-state harmonic current

$$i_{h\,ss} = \frac{1}{l_\sigma} \int u_{h\,ss}\, dt,$$ (7.6)

to the respective steady-state harmonic voltage

$$u_{h\,ss} = u_{ss} - u_1.$$ (7.7)

The definition of the dynamic modulation error (7.2) converts to

$$d_i = \frac{1}{l_\sigma} \int \left(u_{h\,ss} - u_h \right) dt.$$ (7.8)

with reference to (7.5) and (7.6).

7.2 Trajectory of the stator flux harmonics

The measurement of the dynamic modulation error according to (7.8) has been successfully implemented [8]. It requires a fast identification of the total leakage inductance of the drive motor l_σ, which is load-dependent with some machines [9]. The increasing stator current saturates the leakage paths of the stator iron which, in turn, makes the value of the leakage inductance decrease.

The oscillograms Fig. 7.2 were recorded from the low-voltage machine of the setup; the machine parameters are given in Table 2 of the Appendix. Fig. 7.2(a) shows the steady-state current trajectory at no-load operation. The pulse number $N = 8$ and the modulation index $m = 0.47$. The fundamental current magnitude $I_{1,NL} = 22$ A and the peak value of the current harmonics $I_{h,NL} = 5.5$ A; their trajectory is shown in Fig. 7.2(b). Applying full load to the drive system increases the fundamental of the stator current, $I_{1,FL} = 60$ A, Fig. 7.2(c). The leakage inductance value decreases; the peak value of the current harmonics at full load $I_{h,FL} = 10$ A, Fig. 7.2(d), is higher than in the no-load condition, Fig. 7.2(b), due to the variation of the leakage inductance of the specific machine. Subscripts $_{NL}$ and $_{FL}$ relate to operation at no-load and full load, respectively.

Contrary to the harmonic current trajectory, the trajectory of the stator flux harmonics

$$\psi_h = \int u_h\, dt$$ (7.9)

is independent of the leakage inductance. Comparing (7.9) with (7.5), it is evident that the *shape* of the trajectory of the stator flux harmonics

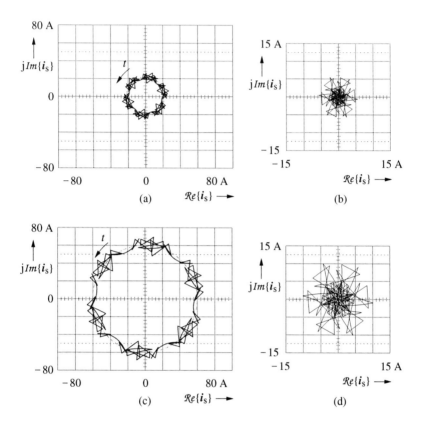

Fig. 7.2 Steady-state stator current trajectories at $N = 8$ and $m = 0.47$ shown in stationary coordinates; (a) stator current trajectory at no-load operation, (b) steady-state current harmonics at no-load, (c) stator current at full load, (d) steady-state curent harmonics at full load.

ψ_h is identical to i_h. However, the estimation of the leakage inductance is not required in order to evaluate the magnitude of the space vector of the stator flux harmonics ψ_h. The proportionality

$$\psi_h = l_\sigma i_h \tag{7.10}$$

allows expressing the dynamic modulation error (7.2) as

$$d = \Psi_{hss} - \Psi_h. \tag{7.11}$$

The error d, without subscript, is thus described as the deviation of the actual stator flux harmonic trajectory from the trajectory of the steady-state harmonic flux.

7.3 Estimation of the dynamic modulation error in real-time

The space vector Ψ_1 of the fundamental stator flux can be included in the right-hand part of (7.11),

$$d = \left(\Psi_1 + \Psi_{hss}\right) - \left(\Psi_1 + \Psi_h\right), \tag{7.12}$$

which then converts to

$$d = \Psi_{ss} - \Psi_s, \tag{7.13}$$

where Ψ_{ss} is the total steady-state stator flux and Ψ_s is the space vector of the stator flux linkage. The estimation of the dynamic modulation error d is preferably based on expression (7.13):

- Acquisition of the space vector of the steady-state stator flux Ψ_{ss} is possible by performing direct integration of the steady-state voltage u_{ss} as obtained from the employed pulse pattern $P(m, N)$.
- Estimation of the space vector of the actual stator flux linkage Ψ_s can be realized by utilizing the measured stator voltage u_s and the stator current i_s as inputs in a model of the stator winding of the induction machine.

The estimated value of the dynamic modulation error

$$\hat{d} = \Psi_{ss} - \hat{\psi}_s \tag{7.14}$$

results from expression (7.13) by substituting the space vector of the stator flux linkage Ψ_s by its estimated value $\hat{\psi}_s$, [5]. The estimation (7.14) is visualized in Fig. 7.3. The signal flow graph builds around the optimal modulator Fig. 5.6, page 65, by adding an estimator for the dynamic modulation error and a model of the stator winding.

7.3.1 Space vector of the steady-state stator flux

To compute Ψ_{ss} in (7.14), the steady-state stator voltage waveform $u_{ss}(t)$ is reconstructed from the set of optimal switching angles $P(m, N)$

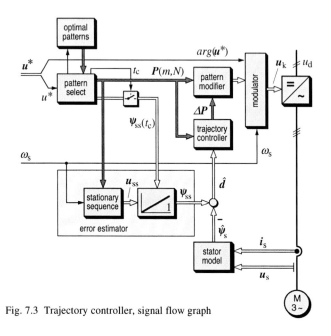

Fig. 7.3 Trajectory controller, signal flow graph

in actual use, and the value ω_s of the stator frequency. Operation in quasi steady-state is assumed; the stator frequency is constant in good approximation. The stored optimal patterns $P(m, N)$ cover the reduced range $0 \leq \alpha \leq \pi/2$, where α is the phase angle of the fundamental component $u_1(\alpha) = \hat{u}_1 \sin\alpha$, with $\hat{u}_1$ being the peak value of the fundamental voltage. We have therefore

in interval I, $0 \leq \alpha \leq \pi/2$: $\qquad u_{ss}^{I}(\alpha) = f\{P(m, N)\}$ $\qquad$ (7.15a)

The remaining portions of a full fundamental period are determined using the conditions for quarter-wave symmetry (5.3), page 60,

in interval II, $\pi/2 \leq \alpha \leq \pi$: $\qquad u_{ss}^{II}(\alpha) = u_{ss}^{I}(\pi - \alpha)$ $\qquad$ (7.15b)

and for half-wave symmetry

in interval III, $\pi \leq \alpha \leq 2\pi$: $\qquad u_{ss}^{III}(\alpha) = -u_{ss}^{I,II}(\alpha - \pi)$ $\qquad$ (7.15c)

The respective functions $u_{ss}(\alpha)$ are used to reconstruct the optimal

84

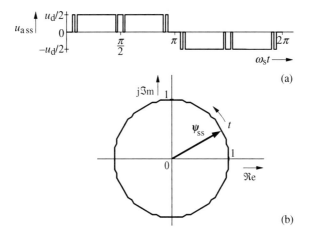

Fig. 7.4 Synchronous optimal pulsewidth modulation; (a) stator voltage wave-form of phase a, (b) trajectory $\Psi_{ss}(t)$ of the steady-state stator flux linkage vector; $m = 0.8$ and $N = 5$.

steady-state waveform $u_{ss}(t)$. This waveform is required in order to determine the steady-state trajectory of the stator flux linkage vector. The reconstruction starts with the initial value $\Psi_{ss}(t_c)$ at the time instant t_c at which the transient is commanded. The pulse pattern must then change. A new pattern $P(m, N)$ is then selected to control the system at $t > t_c$. The steady-state stator flux trajectory for the next interval is therefore

$$\Psi_{ss}(t) = \int_{t_c}^{t} u_{ss}(t)\, dt + \Psi_{ss}(t_c) .\tag{7.16}$$

An illustration of the steady-state trajectory $\Psi_{ss}(t)$ and the originating optimal stator voltage waveform of phase a is given in Fig. 7.4.

The initial value $\Psi_{ss}(t_c)$ in (7.16) is determined in two steps. First, the steady-state trajectory (7.16), which associates to the steady-state waveform $u_{ss}(t)$ as reconstructed from the pulse pattern $P(m, N)$, is integrated over one fundamental period T_1:

$$\int_{0}^{T_1} \Psi_{ss}(t_1)\, dt_1 = \int_{0}^{T_1} \left(\int_{t_c}^{t} u_{ss}(t)\, dt \right) dt_1 + \int_{0}^{T_1} \Psi_{ss}(t_c)\, dt_1 .\tag{7.17}$$

85

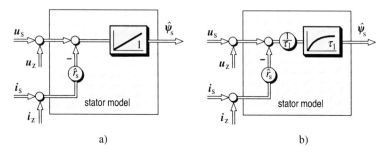

a)   b)

Fig. 7.5 Estimation of stator flux linkage $\hat{\psi}_s$ in stationary coordinates; (a) ideal integrator, (b) first-order element. The space vectors u_z and i_z represent disturbances in the respective voltage and currents inputs of the model.

The left-hand part of (7.17) is equal to zero, since the periodic pulse pattern contributes no dc term in the steady-state trajectory $\psi_{ss}(t)$,

$$0 = \int_0^{T_1}\left(\int_{t_c}^t u_{ss}(t)\,dt\right)dt_1 + T_1 \cdot \psi_{ss}(t_c). \qquad (7.18a)$$

The initial value $\psi_{ss}(t_c)$ in (7.16) then follows as

$$\psi_{ss}(t_c) = -\frac{1}{T_1}\int_0^{T_1}\left(\int_{t_c}^t u_{ss}(t)\,dt\right)dt_1. \qquad (7.18b)$$

7.3.2 Estimation of the stator flux space vector

A first attempt to estimate the space vector of the stator flux linkage is based on the integration of the stator voltage equation (2.3), page 8,

$$\hat{\psi}_s = \int (u_s - \hat{r}_s i_s)\,dt, \qquad (7.19)$$

where $\hat{r}_s$ is the estimated stator resistance. The measured stator current i_s and stator voltage u_s are inputs to the stator model (7.19). This model is visualized in the signal flow graph Fig. 7.5(a). The space vectors are referred to in the stationary coordinate system.

Expression (7.19) is an open integration: there exists no feedback from the output of this structure to its input. Disturbances associated to the stator voltage and the stator current are represented by the respective space vectors u_z and i_z in Fig. 7.5(a): they are superimposed to the in-

86

put of the stator model. The disturbance vector u_z is attributed to the nonlinear characteristics of the PWM inverter, distortion errors of the pulsewidth modulator, and distortions induced by the dead-time effect. Drift and dc offset in the current measurement, erroneous discretization of the measured current signals, and unbalanced gains of the current action channels are represented by the vector i_z. Any dc error component at the input of the open integrator Fig. 7.5(a) will accumulate resulting to a runaway of the output signal $\hat{\psi}_s$; the estimation is erroneous.

To ease the problem, a negative feedback can be added to the integrator in order to stabilize its output [37, 38]. An equivalent representation of such structure is the first-order element Fig. 7.5(b). It converts the open integrator (7.19) to

$$\tau_1 \frac{d\hat{\psi}_s}{d\tau} + \hat{\psi}_s = \tau_1(u_s - \hat{r}_s i_s), \qquad (7.20)$$

where $1/\tau_1$ is the corner frequency of the first-order element. Such model behaves as an integrator only when the frequency of the input signals, i.e. the stator frequency ω_s, is much higher than $1/\tau_1$. As the stator frequency reduces, the integration gain also reduces and the desired 90° phase shift of the output with respect to the input does not hold. The behavior of (7.20) is then that of a low-pass filter in the low-frequency area. Erroneous values are estimated for the magnitude and of the phase angle of the stator flux linkage.

A further source of inaccuracy of the model Fig. 7.5(b) is the value of the stator resistance: it increases with temperature and may vary by more than 100% during operation. An error in the estimated value $\hat{r}_s$ results to a faulty input $\hat{r}_s i_s$. The voltage drop on the stator resistance $\hat{r}_s i_s$ dominates the integrator output in the low-frequency area, since the amplitude of the stator voltage reduces.

More sophisticated methods are required to eliminate the bandwidth restrictions of the model Fig. 7.5(b). Compensation of the inverter nonlinearities, and estimation of the dc offset and of the stator resistance in real-time have been shown to allow accurate estimation of the stator flux linkage at all values of the stator frequency including standstill [39, 40].

Without considering such techniques, the stator model Fig. 7.5(b) has proved robust in the stator frequency range above 0.06 pu [37], where 1 pu corresponds to 50 Hz. Adjusting the corner frequency $1/\tau_1$ accordingly stabilizes the integrator output.

The performance of the model Fig. 7.5(b) is sufficient for the purposes of the present work since the estimated value of the stator flux linkage vector $\hat{\boldsymbol{\psi}}_{\mathrm{s}}$ is only required in the frequency range above 0.3 pu where operation at synchronous optimal modulation is assumed, as shown in Fig. 5.5, page 64.

In addition to this, an estimation error in the value of the stator resistance is negligible since

- the voltage drop on the stator resistance is insignificant in the stator frequency range above 0.3 pu as compared with the induced voltage u_{i}; the stator voltage equals u_{i} in good approximation.
- Other than in low-voltage machines, the stator resistance of medium-voltage machines is very small. For example, the stator resistance of the machine used in the industrial implementation, Section 2.3, is only 69.7 mΩ, as stated in Table 1 of the Appendix.

7.4   Deadbeat control of the dynamic modulation error

To allow quasi steady-state operation of the machine drive at synchronous optimal modulation, detrimental effects of harmonic excursion need to be eliminated. Closed-loop control of the stator flux linkage vector is required.

The optimal steady-state trajectory $\boldsymbol{\psi}_{\mathrm{ss}}$ serves as a target. Deviations of the stator flux linkage vector $\hat{\boldsymbol{\psi}}_{\mathrm{s}}$ from the target trajectory are interpreted as volt-second errors as characterized by the dynamic modulation error $\hat{\boldsymbol{d}}$. This necessitates a modification of the switching angles of the modulator. The volt-second input to the machine is optimized also when then operating point changes; the trajectory controller performs a pattern optimization in real-time.

A strategy to minimize the dynamic modulation error follows two steps: (i) modify any new pulse pattern prior to its use, based on a *prediction* of the dynamic modulation error, and subsequently (ii) eliminate, to the extent possible, the dynamic modulation error during the subsequent sampling interval based on its *estimation*.

7.4.1 Pattern modification

The trajectory controller modifies the actual pulse pattern $P(m, N)$ such that $\hat{\boldsymbol{d}}$ is minimized. The modifying signal is denoted as ΔP in Fig. 7.3. The controller forces the estimated stator flux vector $\hat{\boldsymbol{\psi}}_{\mathrm{s}}$ to track the target trajectory $\boldsymbol{\psi}_{\mathrm{ss}}$ through the control input ΔP.

The compensation of the dynamic modulation error $\hat{\boldsymbol{d}}$ is done consid-

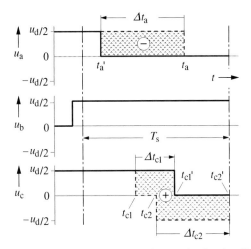

Fig. 7.6 Waveforms of the three phase voltages, showing the effect of error compensation; the dotted lines represent the uncompensated patterns, the solid lines the compensated patterns. The circled sign symbols indicate the polarity of volt-second changes.

ering the switching transitions of the original optimal pulse sequence within the sampling interval $T_s(k)$ under consideration, Fig. 7.6. The following rules can be established, [5]:

- Switching transitions to a more positive potential are characterized by the variable $s = +1$. These are those in which a phase potential changes from $-u_d/2$ to 0, or from 0 to $+u_d/2$, where u_d is the dc link voltage. Delaying such transition by a displacement $\Delta t > 0$ reduces the volt-second contribution of that phase; the contribution increases when the transition is advanced, $\Delta t < 0$.

- Switching transitions to a more negative potential are characterized by $s = -1$. These are those in which a phase potential changes from $+u_d/2$ to 0, or from 0 to $-u_d/2$. Delaying such transition by a displacement $\Delta t > 0$ increases the volt-second contribution of that phase; the contribution decreases when the transition is advanced, $\Delta t < 0$.

- The absence of switching transition in a sampling interval is marked by $s = 0$.

Depending on the particular optimal waveforms of the phase voltages within the sampling interval $T_s(k)$, there may be one or more, and generally n transition steps per phase. Each of these is characterized either by a positive transition, where $s = 1$, or a negative transition, $s = -1$.

Considering phase a, the time displacements Δt_{ai} of the existing n transitions within the sampling interval $T_s(k)$ change the dynamic modulation error by

$$\Delta d_a = -\frac{1}{3} u_d \sum_{i=1}^{n} s_{ai} \Delta t_{ai}. \qquad (7.21)$$

It is assumed that modification (7.21) is executed in the controlling microprocessor during sampling interval $T_s(k)$; the modification was then programmed as $\Delta d_a(k-1)$ in sampling interval $T_s(k-1)$. While the phase-a component $\hat{d}_a(k)$ of the dynamic modulation error is sampled at the beginning of interval $T_s(k)$, the modification $\Delta d_a(k-1)$, computed in the foregoing time interval, has not yet made an effect. However, it will be effective at the end of interval $T_s(k)$. The modification computed during interval $T_s(k)$ is therefore

$$\Delta d_a(k) = -\left(\hat{d}_a(k) - \Delta d_a(k-1) \right). \qquad (7.22)$$

The right-hand side of this equation comes with a minus sign since a required modification is the opposite of an existing error. Thus we obtain from (7.21) and (7.22)

$$\Delta t_{ai} = \frac{3}{u_d} \frac{1}{s_{ai}} \left[\hat{d}(k) - \Delta d(k-1) \right] \circ \mathbf{1} \qquad (7.23a)$$

for any of the $i \in 1 \ldots n$ transition steps, where $\mathbf{1}$ is the unity vector in the direction of phase axis a. The internal vector product on the right hand side of (7.23a) represents the a-phase component of pattern modification in interval $T_s(k)$.

Similarly, the displacements of the respective transition steps in phase b and phase c are

$$\Delta t_{bi}(k) = \frac{3}{u_d} \frac{1}{s_{bi}} \left[\hat{d}(k) - \Delta d(k-1) \right] \circ a \qquad (7.23b)$$

$$\Delta t_{ci}(k) = \frac{3}{u_d} \frac{1}{s_{ci}} \left[\hat{d}(k) - \Delta d(k-1) \right] \circ a^2. \qquad (7.23c)$$

The unity vectors a and a^2 in the directions of the respective phase axis *b* and *c* are shown in the vector diagram Fig. 7.7.

Given the complex nature of displacement vector of the dynamic modulation error Δd, an expansion in terms of its phase components holds:

$$\Delta d = \frac{2}{3}(1 \cdot \Delta d_a + a \Delta d_b + a^2 \Delta d_c). \qquad (7.24)$$

The displacement vector resulting from all manipulations (7.23) during $T_s(k)$ yields

$$\Delta d = -\frac{1}{6} u_d \sum_{i=1}^{n} [(2 s_{ai} \Delta t_{ai} - s_{bi} \Delta t_{bi} - s_{ci} \Delta t_{ci}) \\ + j \sqrt{3} (s_{bi} \Delta t_{bi} + s_{ci} \Delta t_{ci})] \qquad (7.25)$$

The waveforms Fig. 7.6 give an example of how the manipulation of the transient steps within a particular sampling interval $T_s(k)$ is performed. The resulting effect is illustrated in Fig. 7.7. This figure applies for one sampling interval. An error vector $\hat{d}(k)$ exists in the beginning of sampling interval $T_s(k)$. It is the input variable for computing the time displacements (7.23). As time progresses within the current sampling interval, the error vector displaces along the dotted trajectory $\hat{d}(t)$, finally reaching the origin of the complex plane, $\hat{d}(t) = 0$, in this example. The respective time marks on the trajectory correspond to those in the timing diagram Fig. 7.6.

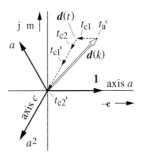

Fig. 7.7 Vector diagram showing the trajectory $d(t)$ of the modulation error during compensation

7.4.2 Properties of stator flux trajectory tracking

The compensation of the dynamic modulation error as part of the stator flux trajectory tracking scheme Fig. 7.3 is performed in a closed loop. The actual stator flux trajectory is reconstructed using the stator model (7.20) based on measured stator currents and stator voltages. The control system does not only handle the adaptation of the optimal steady-state pulse patterns to particular changes of the operating point that may eventually occur. It is also the imperfections of the inverter that are inherently compensated by closed loop control [9]. These are caused by the dead time effect and by the threshold voltages of the power semiconductor devices [41].

A change of the pulse pattern $P(m,N)$, caused by a change of either the modulation index m or of the pulse number N, is the principal cause for the event of a dynamic modulation error. Such error can be predicted *before* the newly selected pulse pattern becomes active. To this end, the difference is taken between the respective steady-state flux linkage vectors of the incoming and the outgoing pulse pattern

$$\tilde{d}(t_c) = \Psi_{ss}(t_c^+) - \Psi_{ss}(t_c^-) \qquad (7.26)$$

where is t_c the sampling instant at which the change of the pulse pattern is detected. The *predicted* error $\tilde{d}(t_c)$ indicates how the pulse pattern must be modified in order to minimize an existing modulation error. Furthermore, $\Psi_{ss}(t_c^+)$ is the initial condition for the prediction of the optimal trajectory of the stator flux linkage vector Ψ_{ss} for the time interval that starts at t_c^+. The signal flow graph Fig. 7.3 shows that $\Psi_{ss}(t_c^+)$ is obtained from the selected optimal switching pattern.

There are, however, some limitations that may impede the complete compensation of the dynamic modulation error:

- A displacement of a transition step in a particular phase exerts only an effect on that very phase component of the dynamic modulation error. This is obvious from an inspection of equations (7.23).
- Advancing all positive-going transition steps ($s = +1$) in the pulse pattern of one particular phase either to the very beginning or delaying them until the end, and delaying all negative-going transition steps ($s = -1$) either to the end or advancing them to the beginning of the sampling interval produces the maximum achievable modification within $T_s(k)$ of that very phase component of the dynamic modulation error. In addition, manipulating the patterns of the other phases may provide further reduction since the three phase components are not completely independent from each other. A complete elimination of the error vector is only achieved if the pulse pattern can be modified [5, 8] by the displacement vector

$$\Delta d = -\hat{d}. \qquad (7.27)$$

In such a case, the displacements Δd_a, Δd_b, and Δd_c take effect within the actual sampling interval; the error vector is forced to the origin of the complex plane. The effectiveness, though, is limited by the maximum achievable modification within $T_s(k)$ [9]. A residual error $\Delta d + d \neq 0$ that should finally remain is transferred to the subsequent sampling interval $T_s(k+1)$.

- If no transitions exist in the modulation pattern of a given phase, that particular phase component of the dynamic modulation error can only less effectively be changed by interfering with the patterns of the other two phases. If needed, an additional inverter commutation may be introduced for this phase. The increase of the switching losses thus incurred should not be an issue since the system normally operates at $f_s < f_{s\,max}$. This is the consequence of synchronization between f_s and the fundamental frequency f_1.

In an implementation of synchronous optimal pulsewidth modulation at 200 Hz switching frequency with the width T_s of the sampling interval set to 500 µs, no commutation within T_s in either of the phases occurred in 10% of the samples. A single commutation occurred in 30% of the samples, two commutations in 40%, and three commutations in 20%. This makes the choice of $T_s = 500$ µs to appear appropriate. A shorter interval would leave less margin for an advancement in time of the transition steps, while a larger interval would increase the response time to commanded changes.

7.4.3 Performance of stator flux trajectory tracking

Synchronous optimal pulsewidth modulation and trajectory control

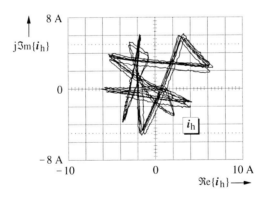

Fig. 7.8 Trajectory of the harmonic current vector in synchronous coordinates; enlarged scale. Operation at $f_1 = 30.5$ Hz, $m = 0.625$, $N = 6$.

93

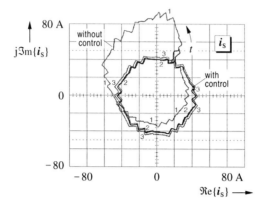

Fig. 7.9   Change of the pulse number from $N = 6$ to $N = 5$ at rated load, trajectory of the stator current vector in stationary coordinates (the numbers 1-3 refer to full revolutions of the fundamental), $f_1 = 30.5$ Hz, $m = 0.625$; gray curve: without trajectory tracking control, black curve: the same process with trajectory tracking engaged.

was implemented in the DSP processor of the drive described in Section 2.3.

Experiments were conducted in the laboratory with the low-voltage prototype inverter, and in the field with the medium-voltage inverter, both feeding their respective induction motors. The machine parameters are given in Tables 1 and 2 of the Appendix.

Although the control system processes the trajectory of the stator flux linkage vector, the trajectories of the stator current vector show the distortions more pronounced and hence were recorded instead for better insight. The typical structures of these trajectories are best viewed at no-load operation where the fundamental current is low. The steady-state trajectory, shown in Fig. 6.6, page 78, at 183 Hz switching frequency, has a six-pulse periodicity. The synchronization between switching frequency and fundamental frequency produces congruent tracks for all fundamental periods. The six periodic intervals per fundamental of the stator current trajectory superimpose when the trajectory is viewed in field coordinates. Fig. 7.8 displays only the harmonic component in an enlarged scale. The dead time effect causes a displacement from the optimal harmonic trajectory whenever one of the phase currents goes

94

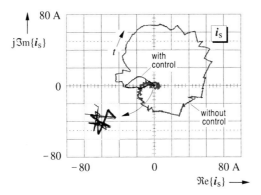

Fig. 7.10 Trajectory of the stator current vector in field coordinates without trajectory tracking control, data as in Fig. 7.8. The dot in the origin represents the final steady-state trajectory. Enlarged inset: The same process with trajectory tracking engaged.

through zero. The deviation is compensated by the trajectory controller. The small jitter is due to noise.

A change of the pulse number can provoke major dynamic modulations errors as shown in the gray curve of the oscillogram Fig. 7.9. To aggravate the situation, the switching distortions get amplified at high-

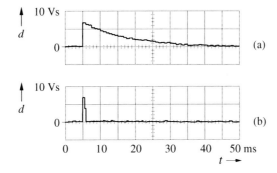

Fig. 7.11 Magnitude of the dynamic modulation error in the process shown in Fig. 7.9; (a) without trajectory tracking, (a) trajectory tracking engaged.

95

er current amplitudes as the leakage inductance of the machine saturates. The large excursion of the stator current vector would have caused an overcurrent trip if the system had less current margin. The harmonic excursion also stimulates eigenoscillations of the machine. These make the current trajectory center only after more than a fundamental period. The added curve in black shows the same process with the trajectory controller engaged. The dynamic modulation error is then predicted and the respective pulse patterns are modified such that the error is minimized.

The transformation to field coordinates in Fig. 7.10 emphasizes the net excursion of the stator current vector. The accumulation around the origin is produced by the decaying eigenoscillations of the machine. The black curve is the response with the trajectory tracking control engaged. The eigenoscillations are almost eliminated.

The region very close to the origin in Fig. 7.10 is magnified 10 times in an inset. Fig. 7.11 shows the error magnitude $\hat{d}(t)$ versus time. The trajectory controller takes two sampling intervals ($2T_s = 1$ ms) for error compensation.

8. Hybrid observer

Overview

The dynamic modulation error serves as a means to measure the undesired excursions that appear at changes of the operating point. Quasi steady-state operation at synchronous optimal modulation is possible by eliminating the error: the trajectory controller detailed in Chapter 7 serves this purpose. Although this goal is achieved, the issue of fast control of the machine torque has not been addressed up to this point.

When the drive is operated at very low switching frequency, the *fundamental* component of the stator current, or stator flux, is employed as feedback signal to acquire high-performance dynamic control. Such signal is inherently obtained as part of the modulation algorithm when carrier based space vector modulation is used [10], Section 3.1.1.

Optimal pulsewidth modulation techniques do not offer a comparable feature. An intrinsic means to extract the fundamental component of the load current is not available; the issue is explained in Section 8.1. An *observer* [42] is used instead; a novel structure is developed in Section 8.2 to acquire the instantaneous values of the fundamental machine quantities. Its performance is analyzed and compared with a classical full-order observer [43] in Section 8.3. Experimental results are provided.

8.1   Space vector of the fundamental stator current

The signal flow graph Fig. 8.1 shows a conventional closed-loop control structure for induction machine drives; operation at space vector modulation is assumed. A current controller provides the reference voltage vector u^*; this is the input to the pulsewidth modulator that generates the switching state vectors u_k to control the power inverter. The stator current reference i_s^* is generated by superimposed controls, e.g. operated at field orientation, Section 4.2. An estimated value of the rotor field angle $\hat{\delta}$ is used to transform variables from stationary coordinates to field coordinates and vice versa as shown in Fig. 8.1. The feedback signal of the control loop is the space vector of the fundamental stator current i_1: it is only the fundamental component of the total stator current i_s that is important for torque control. The fundamental current is sampled at twice the switching frequency $2f_s$, as explained in Section 4.1.

Operation at synchronous optimal modulation does not provide this

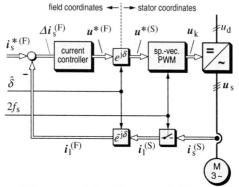

Fig. 8.1 Signal flow graph of closed-loop control of the stator current space
vector; operation at space vector pulsewidth modulation.

benefit; the stator current fundamental cannot be directly sampled. The
problem is visualized in Fig. 8.2. Steady-state harmonic trajectories are
displayed over one sixth of a fundamental period at different values of
modulation index m and pulse number N. It is observed that the trajectories do not necessarily cross the origin of the complex plane: as a consequence, the fundamental current i_1 cannot be sampled at specific time
instants; this is a drawback of synchronous optimal modulation.

High-performance torque control is therefore difficult to obtain. Ex-
tracting the fundamental current by filters produces a signal delay and
thus deteriorates the dynamic performance. A strategy that estimates
the instantaneous fundamental component is therefore desirable. It uses

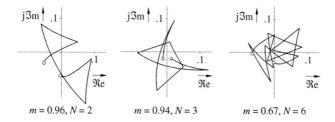

$m = 0.96, N = 2$ $m = 0.94, N = 3$ $m = 0.67, N = 6$

Fig. 8.2 Trajectories of the steady-state harmonic current $i_{h\,ss}$; operation at
synchronous optimal modulation.

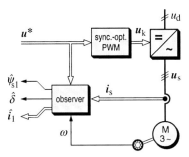

Fig. 8.3 Estimation of instantaneous fundamental quantities by an observer. The estimated values of fundamental stator flux linkage $\hat{\psi}_{s1}$ and fundamental stator current $\hat{i}_1$ are referred to either in the stationary or the field coordinate system; a further output of the observer is the estimated field angle $\hat{\delta}$.

the undistorted reference voltage vector u^* of the synchronous optimal pulsewidth modulator, which is a signal without harmonic content, in combination with an observer for that purpose. The basic structure is illustrated in Fig. 8.3. The observer estimates the fundamental content of either the stator current or of the stator flux linkage vector. It also estimates the rotor field angle $\hat{\delta}$ which is required for the coordinate system transformations, Fig. 8.1. In this work, two different observer structures are analyzed for their suitability.

8.2 Observer structures

8.2.1 Full-order observer

A classical observer structure is derived from the machine equations

$$\tau_\sigma' \frac{di_s}{d\tau} + i_s = \frac{k_r}{\tau_r r_\sigma} (1 - j\omega\tau_r)\psi_r + \frac{1}{r_\sigma} u_s \qquad (8.1a)$$

$$\tau_r \frac{d\psi_r}{d\tau} + \psi_r = j\omega\tau_r\psi_r + l_m i_s, \qquad (8.1b)$$

which are defined in stationary coordinates with the stator current vector i_s and the rotor flux linkage vector ψ_r as state variables. They result from the system equations (2.6), page 11, which are expressed in field coordinates by multiplying them with the unity vector rotator $exp(j\omega_s t)$ for transformation to the stationary coordinate system. The

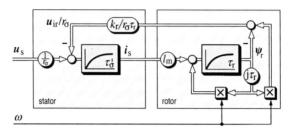

Fig. 8.4 Signal flow graph of the induction motor in stationary coodinates; u_{ir} is the space vector of the rotor induced voltage.

machine model (8.1) is visualized in the signal flow graph Fig. 8.4, [12].

Transforming (8.1) from the continuous-time domain to the frequency domain yields the transfer function of the machine

$$\tilde{I}_s(s) = \frac{1/r_\sigma(\tau_r s + 1 - j\omega\tau_r)\,\tilde{U}_s(s)}{\tau_\sigma' \tau_r s^2 + (\tau_\sigma' + \tau_r - j\omega\tau_\sigma'\tau_r)s + (1 - k_1)(1 - j\omega\tau_r)} \qquad (8.2)$$

where a tilde (~) marks the Laplace transform of a variable and $k_1 = k_r l_m/(r_\sigma \tau_r)$.

To construct a full-order observer for the fundamental machine quantities from (8.1), the harmonic-free reference voltage vector u^* of the synchronous optimal pulsewidth modulator in Fig. 8.3 is used to replace the stator voltage vector u_s. Two error correction terms comprising the tensors $G_s(\omega)$ in the stator equation and $G_r(\omega)$ in the rotor equation are added. This yields

$$\tau_\sigma' \frac{d\hat{i}_1}{d\tau} + \hat{i}_1 = \frac{k_r}{\tau_r r_\sigma}(1 - j\omega\tau_r)\Psi_r + \frac{1}{r_\sigma}u^* + G_s(\omega)\Delta i_s \qquad (8.3a)$$

$$\tau_r \frac{d\Psi_r}{d\tau} + \Psi_r = j\omega\tau_r\Psi_r + l_m(i_s + G_r(\omega)\Delta i_s), \qquad (8.3b)$$

where $\Delta i_s = i_s - \hat{i}_1$ is the error of the estimated fundamental stator current vector $\hat{i}_s$. The signal flow graph of the full-order observer is shown in Fig. 8.5. It is described by the transfer function

$$\tilde{I}_1(s) = \frac{N(s)}{D(s)}, \qquad (8.4a)$$

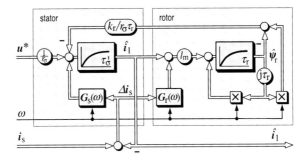

Fig. 8.5 Full-order observer for the estimation of instantaneous fundamental state variables

where

$$N(s) = 1/r_\sigma \cdot (\tau_r s + 1 - \mathrm{j}\omega\tau_r) \cdot \tilde{U}^*(s)$$
$$+ [G_s \tau_r s + (G_s + k_1 G_r)(1 - \mathrm{j}\omega\tau_r)] \cdot \tilde{I}_s(s), \tag{8.4b}$$

and

$$D(s) = \tau_\sigma' \tau_r s^2 + [\tau_\sigma' + (1 + G_s)\tau_r - \mathrm{j}\omega\tau_\sigma' \tau_r]s$$
$$+ [(1 + G_s) - k_1(1 - G_r)](1 - \mathrm{j}\omega\tau_r) \tag{8.4c}$$

The error correction signals fed through the two tensors $G_s(\omega)$ and $G_r(\omega)$ refer to the actual system state represented by i_s. They make the observer robust against disturbances. A design procedure for $G_s(\omega)$ and $G_r(\omega)$ based on the respective system eigenvalues is proposed in [43]. Given the eigenvalues $\lambda_{1,2\,\mathrm{obs}}$ of the observer and $\lambda_{1,2\,\mathrm{mach}}$ of the machine, the design rule is

$$\lambda_{1,2\,\mathrm{obs}} = \varepsilon \cdot \lambda_{1,2\,\mathrm{mach}}, \tag{8.5}$$

where $\varepsilon > 1$ is a real constant.

Applying (8.5) to the respective solutions of the characteristic equations of the machine (8.2) and of the observer (8.4) leads to

$$\tau_\sigma' + (1 + G_s)\tau_r - \mathrm{j}\omega\tau_\sigma' \tau_r = \varepsilon(\tau_\sigma' + \tau_r - \mathrm{j}\omega\tau_\sigma' \tau_r) \tag{8.6a}$$

$$(1 + G_s) - k_1(1 + G_r) = \varepsilon^2(1 - k_1) \tag{8.6b}$$

from which the respective tensors

$$G_s(\omega) = (\varepsilon - 1)\left(\frac{\tau_\sigma{}'}{\tau_r} + 1 - j\omega\tau_\sigma{}'\right) \tag{8.7a}$$

$$G_r(\omega) = \frac{1}{k_1}\left[(\varepsilon^2 - 1)(1 - k_1) - (\varepsilon - 1)\left(\frac{\tau_\sigma{}'}{\tau_r} + 1 - j\omega\tau_\sigma{}'\right)\right] \tag{8.7b}$$

are obtained. The output of the full-order observer is the estimated instantaneous fundamental component $\hat{i}_1$ of the stator current vector.

8.2.2 Hybrid observer

An alternative observer is established in a hybrid structure as the combination of the stator model in stationary coordinates

$$\tau_s{}'\frac{d\hat{\psi}_{s1}{}^{(S)}}{d\tau} + \hat{\psi}_{s1}{}^{(S)} = k_r\hat{\psi}_r{}^{(S)} + \tau_s{}'u^{*(S)} + G_s(\omega)\Delta\psi_s{}^{(S)} \tag{8.8a}$$

and the rotor model in field coordinates

$$\tau_r\frac{d\hat{\psi}_r{}^{(F)}}{d\tau} + \hat{\psi}_r{}^{(F)} = -j\hat{\omega}_r\tau_r\hat{\psi}_r{}^{(F)} + l_m i_s{}^{(F)} \tag{8.8b}$$

where $\tau_s{}' = \sigma l_s/r_s$ is a transient time constant of the stator and $\Delta\psi_s{}^{(S)} = \hat{\psi}_s - \hat{\psi}_{s1}$ is the estimated stator flux error. The superscripts indicate the respective coordinate system. The stator model (8.8a) is derived from the signal flow graph Fig. 2.5, page 11, where the stator flux linkage ψ_s and rotor flux linkage ψ_r are selected as complex state variables. The stator equation (2.9a), page 12, is multiplied with the unity vector rotator $exp(j\omega_s t)$ to transform it to the stationary coordinate system. The stator flux space vector is replaced by its respective estimated fundamental value $\hat{\psi}_{s1}$. Finally, the error correction term comprising the tensor $G_s(\omega)$ is added to yield (8.8a).

The signal flow graph of the hybrid observer [5] is shown in Fig. 8.6. The state variable of the stator model is the fundamental component $\hat{\psi}_{s1}$ of the stator flux linkage vector. It is derived from the reference voltage vector u^* as the input signal and the actual state of the machine as represented by the estimated rotor flux linkage vector $\hat{\psi}_r$.

The rotor model operates at field orientation, Section 4.2, and hence $\hat{\psi}_{rq} = 0$ holds. Equation (8.8b) then becomes

$$\tau_r\frac{d\hat{\psi}_{rd}}{d\tau} + \hat{\psi}_{rd} = l_m i_d \tag{8.9a}$$

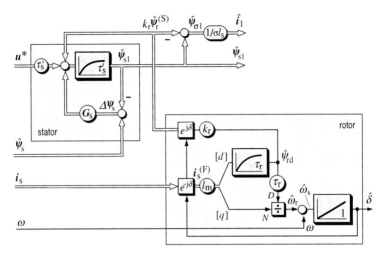

Fig. 8.6 Hybrid observer for the estimation of instantaneous fundamental
state variables

$$\hat{\omega}_r = \frac{l_m i_q}{\tau_r \hat{\psi}_{rd}} . \qquad (8.9b)$$

The condition for rotor field orientation (8.9b) determines the slip
frequency $\hat{\omega}_r$. It is added to the measured angular velocity ω to yield
the stator frequency, $\hat{\omega}_s = \omega + \hat{\omega}_r$. This signal is subsequently integrated to identify the field angle

$$\hat{\delta} = \int \hat{\omega}_s d\tau . \qquad (8.9c)$$

The first-order differential equation (8.9a) allows estimating the real
component of the rotor flux linkage vector $\hat{\psi}_{rd}$ from the respective stator current scalar component i_d. The rotor flux linkage vector $\hat{\psi}_r^{(S)}$ in
(8.8a) is obtained by transforming $\hat{\psi}_{rd}$ from (8.9a) to stationary coordinates using the estimated rotor field angle $\hat{\delta}$ given by (8.9c). The resulting signal $k_r \hat{\psi}_r^{(S)}$ is an input to the stator model.

A further input to the stator model in Fig. 8.6 is the estimated stator
flux vector $\hat{\psi}_s$, obtained from the output of the estimator Fig. 7.5(b),
page 86. It was originally used to define the dynamic modulation error
(7.14), page 83. Here, the space vector $\hat{\psi}_s$ serves a further purpose: it

103

forms the stator flux error

$$\Delta \hat{\psi}_s^{(S)} = \hat{\psi}_s - \hat{\psi}_{s1} \qquad (8.10)$$

from which the error correction signal $G_s \cdot \Delta \psi_s^{(S)}$ is computed. The correction factor G_s is chosen as $G_s = g_1 + j0$, i.e. with a zero imaginary part. Such selection is justified by inspecting the stator equation (8.8a) of the hybrid observer: other than in the respective equation (8.3a) of the full-order observer, there exist no imaginary cross-coupling terms in (8.8a).

The instantaneous fundamental component of the stator current vector is estimated at the output of the hybrid observer

$$\hat{i}_1 = \frac{1}{\sigma l_s} \left(\hat{\psi}_{s1} - k_r \hat{\psi}_r \right). \qquad (8.11)$$

It is useful to emphasize the fact that the hybrid observer was designed starting from equations (2.9), page 12, using ψ_s and ψ_r as state variables: its function is to guarantee an accurate estimation of the fundamental stator flux $\hat{\psi}_{s1}$, which is also denoted as a system output in Fig. 8.6. The fundamental current (8.11) is an auxiliary output of the observer: it is only employed to compare the dynamics of the hybrid observer with those of the full-order observer (8.3) under the assumption that the load-dependent leakage inductance of the machine is known.

8.3 Dynamic behavior

8.3.1 Observer dynamics

Both observers were derived from the model equations of the machine. Robustness against disturbances is achieved by means of an additional input signal. The measured stator current vector is best suited for this purpose.

Observer disturbances originate from two sources:

(i) from a mismatch between the respective parameters of the observer and the machine, and

(ii) from the fact that the observer receives input signals that differ from the input signals of the machine.

The observers have the reference voltage vector u^* of the pulsewidth modulator as an input while the actual stator voltages of the machine are delayed in time by the pulsewidth modulator and distorted by the

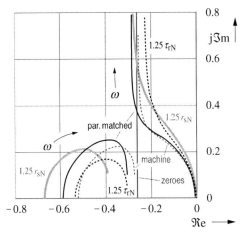

Fig. 8.7 Root loci of the full-order observer with and without parameter errors, $\varepsilon = 1.12$; dotted lines show the root loci of the machine for comparison.

inverter and its nonlinearity [39]. The stator current signal i_s, as an input to the observer, is further delayed against the machine currents owing to time discrete signal sampling.

The observer itself changes its dynamic behavior under the influence of the error correction signals. The signal flow graph Fig. 8.5 of the full-order observer shows that its stator delay element receives the input $G_s(\omega)\cdot\varDelta i_s$, and its rotor delay element receives the input $G_r(\omega)\cdot\varDelta i_s$. Both signals depend on $\hat{i}_1$, which is a state variable of the observer. Hence the two feedback terms, introduced with the objective to make the observer robust, also modify the dynamic behavior of the observer.

This is exemplified in a comparison between the eigenvalues of the observer and of the machine. The eigenvalues are derived from the respective characteristic equations of the observer (8.4) and of the machine (8.2). Fig. 8.7 shows the result. Each system is of second-order and hence characterized by a set of two curves. These are the traces on which the respective two single-complex eigenvalues displace while the angular velocity ω of the rotor varies in the range $\omega = 0 \dots 0.8$.

The eigenvalue curves of the full-order observer are shown in black in Fig. 8.7 with the observer parameters matched to those of the machine. The eigenvalue curves of the machine (dotted gray lines) are shown for comparison. The curves of the observer are displaced with respect to those of the machine to larger magnitude values according to the design rule (8.5).

Note that the system equations, i.e. (8.1), or (8.3), are established in terms of complex state variables. Their Laplace transforms and the respective eigenvalues are then also complex. The eigenvalues do not appear as conjugate complex pairs, rather as single-complex entities. An interpretation of the physical significance of single-complex eigenvalues is given in [44].

Errors in the observer parameters account for further deviations of the observer eigenvalues. Two more sets of eigenvalue curves are included in Fig. 8.7, computed for $r_s = 1.25 \cdot r_{sN}$ (shown by thick light gray lines), and for $\tau_r = 1.25 \cdot \tau_{rN}$ (shown by dashed black lines), respectively. The subscript $_N$ indicates a nominal value. Note that the parameters of the machine may change in a much wider range of up to 100% owing to variations in the winding temperature of the machine. The observer is then in error unless the parameter variations of the machine are traced by on-line identification techniques, and the observer parameters are corrected.

The second source of inaccuracies are different signal delays in the drive system and the observer. Signal delays in a time-discrete sampling system do not reflect in the eigenvalue curves. To investigate their influence, the signal delays are modelled using the structure shown in the signal flow graph Fig. 8.8. The system input is the reference voltage vector u^*. Its value $u^*(n)$ at sampling instant n controls the pulsewidth modulator, which, after a time delay T_s of one sampling interval, creates a set of k switching state vectors $\{u_k(n+1)\}$. These, one after another, control the continuous-time system that represents the inverter and the machine during the subsequent time interval. The system generates the stator current vector $i_s(t)$ as an output. Its sampled value $i_s(n+2)$ is an input to the time-discrete observer. Other inputs to the observer are the reference voltage vector $u^*(n)$, which is not delayed, and the angular mechanical velocity ω of the rotor. The observer thus operates with input signals that represent the system state at different instants of time. To estimate the fundamental content of the sampled current $i_s(n+2)$, which represents the actual system state, the information of $u^*(n)$ is used which is advanced in time by two sampling intervals. It is only two time steps later that this signal will determine the system state.

It should be recalled here that the instantaneous fundamental component of a distorted nonperiodic signal is not uniquely defined. The defi-

106

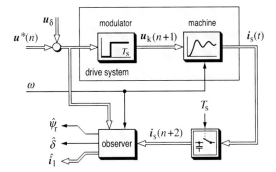

Fig. 8.8 Signal flow graph of the structure used for simulation; u_k are the discrete switching state vectors, u_d is a complex unit Dirac pulse.

nition of the magnitude of the fundamental stator current

$$I_1 = \frac{1}{T_1} \int\limits_{(T_1)} i_s(t) \cdot e^{-j\omega_s t} dt \qquad (8.12)$$

is only valid for periodic signals, $i_{1\,per}(t) = I_1 \cdot exp(j\omega_s t)$. Equation (8.12) requires the signal content of a full fundamental period T_1 as an input, and holds only if the fundamental frequency ω_s does not change. Both conditions do not generally apply in a high-performance drive. The observer tracks the *instantaneous* fundamental current $i_1(t)$: it is the current that the machine would produce if its stator windings were fed by *sinusoidal* voltages of *adjustable* frequency. Such definition includes (8.12), but it is not restricted to operation in steady-state. It is the correct estimation of the instantaneous fundamental current in transient conditions that justifies the proposed method; results are given in Section 8.3.

To summarize, the signal flow graph Fig. 8.8 represents the drive system and the observer under time-discrete control. The structure is used in a simulation to investigate the tracking capability of the observer under transient conditions.

In the simulation, the drive system is assumed to operate with a given value ω of the angular mechanical velocity. The test excitation signal is a voltage in form of a complex Dirac impulse. It is added to the input of the system at the time instant $t = 0$. The voltage impulse $u_\delta = (u_\delta + j0) \cdot \delta(t)$ is defined in alignment with the real axis of the stationary reference frame; $\delta(t)$ is the unit Dirac pulse.

The effect of such excitation is a transient deviation

$$\breve{i}_s = i_s(t) - i_{ss}(t) \tag{8.13}$$

of the stator current vector from its steady-state trajectory $i_{ss}(t)$. This trajectory would have been the output if the drive system had continued in an undisturbed steady-state. The resulting impulse response $\breve{i}_s(t)$ in (8.13) characterizes the dynamic behavior of the observer.

As the decay of an impulse-generated transient is dominated by the transient stator time constant τ_σ', the expected transient lasts only a few milliseconds. It is therefore justified in the model Fig. 8.8 that the mechanical angular velociry ω does not significantly change during that transient. This is particularly true when the impulse aligns with the rotor flux vector.

8.3.2 Performance comparison

8.3.2.1 Simulation results

The full-order observer and the hybrid observer are compared in terms of their respective impulse response trajectories. To study their robustness against parameter errors, $r_{s\,obs} = 1.25 \cdot r_{sN}$ is assumed as an example. The complete set of machine data is listed in Table 2 of the Appendix.

Fig. 8.9 shows that the trajectory of the full-order observer (dashed line) exhibits a deviation from the target trajectory, which is the impulse response trajectory of the actual machine. The latter is represented in this graph by a thick gray line. Varying the speed has little influence on the error trajectories as Fig. 8.10 confirms.

The impulse response trajectory of the hybrid observer (shown by a black line in Fig. 8.9 and 8.10) follows the target trajectory with much better accuracy. The error $\Delta\breve{i}_s(t)$ between the impulse response trajectories of the full-order observer and the target trajectory is displayed in Fig. 8.11(a) as a function of the design parameter ε. Operation at rated speed, $\omega = 1$, and a parameter error $r_{s\,obs} = 1.25 \cdot r_{sN}$ are considered. Similar error trajectories are displayed in Fig. 8.11(b) for the hybrid observer with the feedback gain G_s as a parameter.

The trajectories are identical in the absence of the respective error compensation signal: the error trajectory of the full-order observer at $\varepsilon = 1$ in Fig. 8.11(a) is the same as the error trajectory of the hybrid observer at $G_s = 0$ in Fig. 8.11(b).

The error compensation signals are activated in the following. The set of error trajectories of the full-order observer in Fig. 8.11(a) demonstrate that the average dynamic deviation of $\Delta \breve{i}_s$ assumes a minimum for a certain value of the design parameter ε. However, a more detailed study shows that the design parameter ε should have either a larger or a lower value for better disturbance rejection, depending on the particular parameter that is in error. A similar requirement for ε exists when an observer is used for parameter identification [45].

Error trajectories $\Delta \breve{i}_s(t)$ of both observers with the observer parameters set as $\tau_{r\,obs} = 1.25 \cdot \tau_{rN}$ are shown in Fig. 8.12.

Other than with the full-order observer, the average dynamic deviation of the hybrid observer reduces monotonously while the feedback gain G_s increases. The increasing gain changes the dynamic behavior of the hybrid observer significantly. This can be confirmed by an observation of its eigenvalue curves. They are similar to those of the full-order observer, Fig. 8.7, at low feedback gain. As the gain increases, the path of one of the eigenvalues, λ_1, converts to a straight line, Fig. 8.13. Its imaginary part $\Im\{\lambda_1\} \cong j\omega$. The eigenvalues locate in very close neighborhood to the parallel straight line of the observer zeroes. The zeroes then almost eliminate the influence of these poles. The desired effect is already achieved at moderate values of the feedback gain G_s. Interestingly, these eigenvalues hardly change even when the observer parameters $r_{s\,obs}$ and $\tau_{r\,obs}$ vary within ±25%.

A value $G_s = 3.6$ was chosen for the experiments. The choice is not critical, although an inspection of the signal flow graph Fig. 8.6 shows that high values of G_s would let the harmonic content of i_s appear in the waveform of the estimated fundamental current.

What remains effective is a single-complex pole λ_2, located farther away on the negative real axis. The transfer function of the hybrid observer therefore approximates a fast first-order delay element. Its time constant depends on the feedback gain G_s.

The comparison in Fig. 8.14 of the magnitudes of the two impulse response functions indicate that the error incurred by the full-order observer is essentially a time lag of about $2T_s$. Its delayed response makes the full-order observer less suited for the intended purpose, which is the estimation of the fundamental component of the stator current vector.

8.3.2.2 Experimental results

The experiments to confirm the estimation of the instantaneous fun-

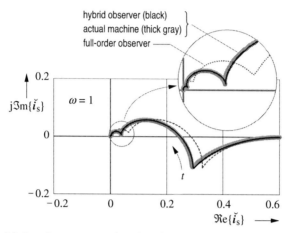

Fig. 8.9 Impulse response trajectories of the full-order observer ($\varepsilon = 1.06$) and of the hybrid observer ($G_s = 3.6$) at $\omega = 1$; observer parameter $r_{s\,obs} = 1.25 \cdot r_{sN}$; impulse response trajectory of the actual machine for comparison.

damental current are conducted using the low-voltage drive of the experimental setup fed from a three-level inverter using synchronous optimal pulsewidth modulation. The machine data are listed in Table 2 of Appendix. The switching frequency varies in a range 180 ... 220 Hz, depending on the operating point. The time-discrete control signals are processed at 2 kHz sampling rate ($T_s = 500$ µs). The trajectory control-

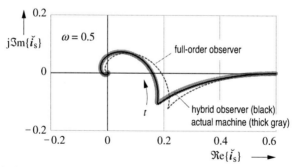

Fig. 8.10 Impulse response trajectories of the full-order observer ($\varepsilon = 1.06$) and of the hybrid observer ($G_s = 3.6$) at $\omega = 0.5$; observer parameter $r_{s\,obs} = 1.25 \cdot r_{sN}$; impulse response trajectory of the actual machine for comparison.

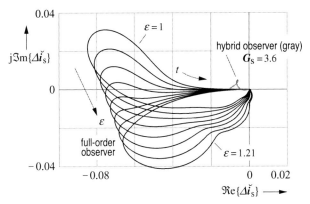

Fig. 8.11(a)   Error of the impulse response trajectories of the full-order observer, ε = parameter; observer parameter $r_{s\ obs} = 1.25 \cdot r_{sN}$; impulse response error of the hybrid observer for comparison.

ler described in Section 7.4, is employed to eliminate the excursion of harmonics at transient operation: the harmonic component of the stator flux linkage and of the stator current follow an optimized trajectory. Fast control of the fundamental stator current is also employed: the re-

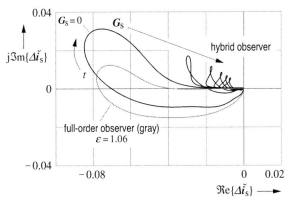

Fig. 8.11(b)   Error of the impulse response trajectories of the hybrid observer, G_s = parameter; observer parameter $r_{s\ obs} = 1.25 \cdot r_{sN}$; impulse response error of the full-order observer with $\varepsilon = 1.06$ for comparison.

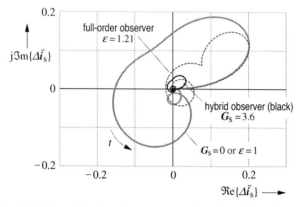

Fig. 8.12 Error of the impulse response trajectories of the full-order and hybrid observer, observer parameter $\tau_{r\,obs} = 1.25 \cdot \tau_{rN}$; impulse response error of the observers without error correction signals ($\varepsilon = 1$, $G_s = 0$) for comparison.

lated method is described in Chapter 8.

Fig. 8.15 demonstrates the extraction of the fundamental component $\hat{i}_{a1}$ from the distorted a-phase current waveform at steady-state. Although the switching frequency is less than 200 Hz, an immediate response is obtained without errors with respect to phase angle and magnitude. The time-discrete signal follows almost exactly the ideal waveform i_{a1} represented in the graph by a gray sinewave.

The corresponding time-discrete trajectory $i_1(n)$ of the estimated fundamental current space vector is shown in Fig. 8.16 with the continuous trajectory $i_s(t)$ of the measured current space vector shown in gray as a reference. Steady-state results as in Fig. 8.15 and Fig. 8.16 can be equally obtained with both types of observers.

The dynamic performance of the two observers is compared in the oscillograms Fig. 8.17 and Fig. 8.18. They display the waveforms of the current components i_d and i_q at rotor field orientation during an acceleration of the machine. The excitation is a large-signal torque step with the stator current limited to $i_{s\,max} = 1.5\,i_{sR}$. The response of the full-order observer has a rise time of 16 ms, and oscillations persist thereafter. Against this, the set point is reached after 3 ms with the hybrid observer at negligible overshoot. Cross-coupling between i_d and i_q is much higher with the full-order observer than with the hybrid observer.

112

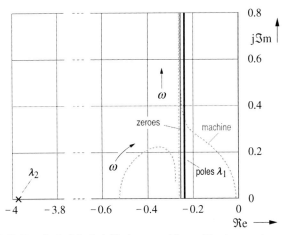

Fig. 8.13 Root loci of the hybrid observer with or without parameter errors, $G_s = 3.6$; the gray dashed lines are the root loci of the machine.

The oscillations with the full-order observer are not due to a wrong setting of the observer or controller parameters. This becomes apparent from recorded small-signal tests. A torque step of 50% rated magnitude in Fig. 8.19 shows the absence of oscillations. Also here is the response sluggish. The hybrid observer performs within two sampling intervals (1 ms) with no overshoot and almost zero cross-coupling, Fig. 8.20.

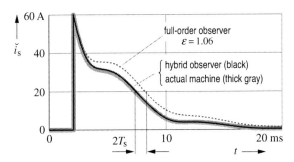

Fig. 8.14 Impulse response waveforms of the full-order and hybrid observers; the thick gray trace is the impulse response waveform of the actual machine for comparison.

113

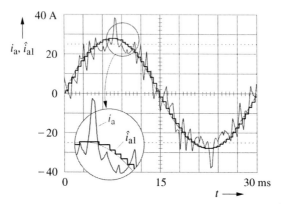

Fig. 8.15 Estimated fundamental component of the phase current $\hat{i}_{a1}$ at 33 Hz fundamental frequency, steady-state operation. The sampling interval is 500 μs; the gray curve shows the continuous waveform of the measured a-phase current, the continuous-time sinusoidal curve is the ideal fundamental component i_{a1}. Signal details are shown in the enlarged inset.

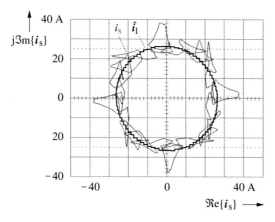

Fig. 8.16 Estimated time-discrete trajectory $\hat{i}_1(n)$ of the fundamental component of the current space vector at steady-state; the gray curve shows the continuous trajectory $i_s(t)$ of the measured current space vector.

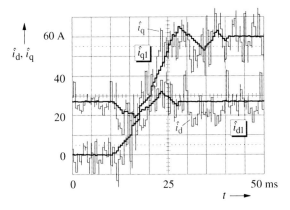

Fig. 8.17 Full-order observer: Large-signal response to a torque step input with the stator current limit set to $i_{s\,max} = 1.5\ i_{sR}$, showing the estimated fundamental current components $\hat{i}_{d1}$ and $\hat{i}_{q1}$ in black. The corresponding sampled signals i_d and i_q are shown in gray.

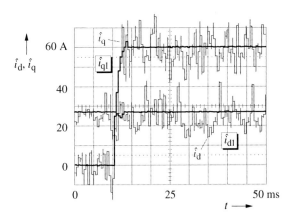

Fig. 8.18 Hybrid observer: Large-signal response to a torque step input with the stator current limit set to $i_{s\,max} = 1.5\ i_{sR}$, showing the estimated fundamental current components $\hat{i}_{d1}$ and $\hat{i}_{q1}$ in black. The corresponding sampled signals i_d and i_q are shown in gray.

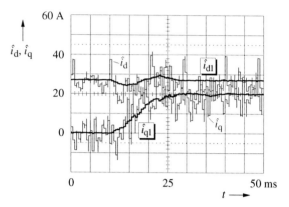

Fig. 8.19 Full-order observer: Small-signal response to a torque step input, showing the estimated fundamental current components $\hat{i}_{d1}$ and $\hat{i}_{q1}$ in black. The corresponding sampled signals i_d and i_q are shown in gray.

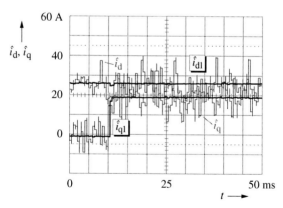

Fig. 8.20 Hybrid observer: Small-signal response to a torque step input, showing the estimated fundamental current components $\hat{i}_{d1}$ and $\hat{i}_{q1}$ in black. The corresponding sampled signals i_d and i_q are shown in gray.

9. High-performance dynamic control by stator flux trajectory tracking

Overview

The investigation in the present Chapter concentrates on the control of *fundamental* machine quantities: the goal is to achieve high-performance torque control with the machine drive operated at synchronous optimal modulation. Instead of relying on the fundamental current, which is dependent on the leakage inductance of the machine (8.11), page 104, the space vector of the fundamental flux is employed as feedback signal in a closed-loop scheme: its accurate estimate is provided by the hybrid observer, Chapter 8.

Employing linear controllers, such as the ones described in Chapter 4, is not possible in combination with synchronous optimal pulsewidth modulation. The controller output tends to adversely interfere with the optimal patterns [9, 46]. This issue is analyzed in Section 9.1.

A novel closed-loop scheme is introduced in this Chapter. It prevents the interference of the control actions with the optimal pulse patterns and enables dynamic decoupling between torque and flux. Section 9.2 introduces the principle of a self-controlled machine that enables the modulator to operate from a quasi steady-state input signal, even when dynamic operation is demanded. This signal is derived from the output of the hybrid observer; the process is detailed in Section 9.3.

A nonlinear controller is employed to achieve high-performance control: its function is based on stator flux trajectory tracking, Sections 9.4 and 9.5. The operation of the trajectory controller amounts to real-time optimization of the switching pattern during transients. Its dynamic performance is evaluated in Section 9.6.

9.1 The steady-state requirement

Closed-loop torque and flux control in high-performance drives is conventionally achieved by controlling the fundamental stator current. Linear current controllers were outlined in Chapter 4. Their application in a system operating at synchronous optimal modulation is not a viable solution: small disturbances and signal noise would provoke correcting actions of the superimposed controllers even at quasi steady-state operation. The controlling input u^* of the modulator would persistently change, thus evoking transitions between different pulse patterns. In con-

117

sequence, dynamic modulation errors would occur in every sampling interval. Although the trajectory controller described in Section 7.4, is designed to counteract such deviations, its complete control potential would be exhausted alone for this purpose. The deviations would then amplify when larger changes of u^* occurred at transient operation. Such situation is undesirable; the reasons are summarized in the following:

- The volt-second error caused by a pattern change can be substantial even when the noise-induced variations of the reference voltage vector u^* are small. This is because the functions $\alpha_k(m, N)$ of the optimal angles are discontinuous, Fig. 5.4 and 5.5, pages 63-64. A small deviation of the modulation index m may lead to large displacements in time of the PWM pulses. A full compensation is then only possible if favored by the particular locations of pulse transitions in the respective sampling interval.

- Since dynamic modulation errors may appear in a rapid sequence, the resulting deviations may accumulate to large magnitudes before the trajectory controller can counteract.

Also the stator frequency signal ω_s would exhibit unnecessary excursions in transient conditions. Being an input to the modulator in Fig. 7.3, page 84, the excursions would cause erroneous conversions of switching angles to time instants. They would also introduce dynamic errors in the target trajectory ψ_{ss}, (7.16), being an input to the error estimator.

It is therefore expedient to maintain the modulator in an approximate steady-state, so as to meet the steady-state condition at which the pattern optimization was performed.

9.2 The self-controlled machine

Ideal steady-state operation is difficult to establish in a drive system. Signal noise and external disturbances inevitably generate minor, but perpetual excursions of the control signals around the operating point. The requirement of near-steady-state operation of the modulator is met in an approach in which the terminal voltages of the machine generate the reference voltage vector u^* for the modulator, [11].

In Fig. 9.1, a *first estimate* of the reference voltage vector $u^{*'}$ is created as a replica of the fundamental component of the stator voltages at the machine terminals

$$u^{*'} = j\hat{\omega}_{sss}\,\hat{\psi}_{s1} + r_s\hat{i}_1. \qquad (9.1)$$

118

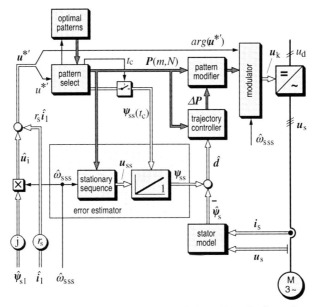

Fig. 9.1   Signal flow graph of the self-controlled machine. The first estimate of the reference voltage $u^{*'}$ is created as a replica of the fundamental stator voltage component $\hat{u}_1$ to select an optimal pulse pattern $P(m, N)$ for inverter control. The steady-state stator frequency $\hat{\omega}_{s\,ss}$ is a further estimated quantity.

In this equation, the fundamental components $\hat{\psi}_{s1}$ of the stator flux linkage vector and $\hat{i}_1$ of the stator current vector, as well as the quasi-steady-state component $\hat{\omega}_{s\,ss}$ of the stator frequency ω_s are estimated quantities; they serve as inputs in the signal flow graph Fig. 9.1. The stator resistance r_s enters as a constant value since its detuning is not significant at $m > 0.3$, as explained in Section 7.3.2.

The proposed concept uses the estimated fundamental component $\hat{u}_1 = u^{*'}$, (9.1), of the machine terminal voltages as the controlling input to the pulsewidth modulator. The first estimate of the reference voltage $u^{*'}$ selects the appropriate pulse pattern $P(m, N)$ that controls the inverter. The steady-state stator frequency $\hat{\omega}_{s\,ss}$ translates the switching angles a_k of the optimized pulse pattern to time instants. Thus, whatever the machine voltages are like with respect to amplitude and frequen-

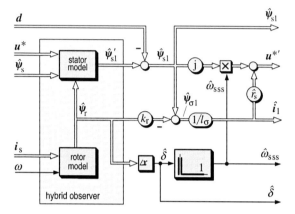

Fig. 9.2 Estimation of the equivalent steady-state quantities. The fundamental quantities of the output are used in (9.1) to establish self-control of the machine. The stator flux fundamental $\hat{\psi}_{s1}$ is employed as feedback signal in Fig. 9.3 to achieve dynamic control. The estimated current fundamental $\hat{i}_1$ is evaluated for monitoring purposes.

cy, they are estimated and reestablished at the machine terminals through the modulator and the inverter.

The machine thus controls itself. The system has no external control input. It would, ideal conditions assumed, indefinitely continue perpetuating any given steady-state operating point. The absence of external control input implies that the drive system is not controllable. Controllability is established by separate means which are discussed in Section 9.4.

To acquire the first estimate of the voltage reference vector $u^{*'}$, the respective values of the stator flux fundamental $\hat{\psi}_{s1}$ and stator current fundamental $\hat{i}_1$ are needed; their estimation is described next.

9.3 Equivalent steady-state quantities

The signal flow graph Fig. 9.2 illustrates the general concept of estimation of fundamental state variables. The hybrid observer described in Section 8.2.2, receives the reference voltage space vector u^* and the measured angular velocity ω as inputs; further inputs are the machine currents i_s. The stator is modelled in stationary coordinates (8.8a) and the rotor in field coordinates (8.8b), page 102.

120

The rotor model estimates the rotor field angle $\hat{\delta}$ to perform the transformations between coordinate systems. It also estimates the rotor flux vector $\hat{\boldsymbol{\psi}}_r$.

The stator model of the hybrid observer receives the voltage reference vector $\boldsymbol{u}^*$ and $\hat{\boldsymbol{\psi}}_r$ as inputs. A further input is the estimated value of the stator flux linkage $\hat{\boldsymbol{\psi}}_s$. It is employed in order to force the observer dynamics to match those of the induction machine. In such way, the observer can account for model parameter mismatch and signal delay. The observer generates a first estimate of the fundamental component $\hat{\boldsymbol{\psi}}'_{s1}$ of the stator flux vector at its output. Since the input $\boldsymbol{u}^*$ does not contain harmonic components, the output components are also free of harmonics.

The trajectory controller in Fig. 9.1 modifies the selected pulse pattern $\boldsymbol{P}(m, N)$ by shifting its precalculated switching angles α_k. The modification process, detailed in Section 7.4, minimizes the dynamic modulation error $\hat{\boldsymbol{d}}$. The inverter is therefore controlled by the optimal pulse pattern $\boldsymbol{P}(m,N)$ that $\boldsymbol{u}^*$ invokes, *minus* the modification $\Delta\boldsymbol{P}(\hat{\boldsymbol{d}})$ that the trajectory controller introduces, Fig. 9.1. Considering the correcting action taken by the trajectory controller, the fundamental stator flux linkage of the machine becomes

$$\hat{\boldsymbol{\psi}}_{s1} = \hat{\boldsymbol{\psi}}'_{s1} - \hat{\boldsymbol{d}}. \tag{9.2}$$

Expression (9.2) is visualized in the estimation scheme Fig. 9.2. A further output of the signal flow graph is the estimated fundamental current $\hat{\boldsymbol{i}}_1$: its value is employed to estimate the voltage drop $r_s\hat{\boldsymbol{i}}_1$ on the stator resistance and for monitoring purposes of the controller performance. The control itself uses the fundamental stator flux as feedback signal. It is described in the next Section.

The fundamental machine quantities allow allocating the space vector $\hat{\boldsymbol{u}}_1 = \boldsymbol{u}^{*'}$ that represents the terminal voltages at the machine. The notation $\boldsymbol{u}^{*'}$ in Fig. 9.2 characterizes the value as a first estimate of the reference voltage vector $\boldsymbol{u}^*$. It is evaluated in real-time by (9.1) to establish self-controlled operation of the machine.

The quasi-steady-state stator frequency signal

$$\hat{\omega}_{sss} = \frac{d\hat{\delta}}{d\tau} = \frac{d}{d\tau}\left(arg(\hat{\boldsymbol{\psi}}_r)\right) \tag{9.3}$$

is derived from the angular velocity of the rotor flux linkage vector. The construction uses the large rotor time constant of the machine τ_r as a filter to obtain a smooth stator frequency signal.

9.4 Dynamic control through pattern modification

The next step is to add a control input to the self-controlled machine, and to superimpose controllers for torque, speed, and rotor flux. Using the reference voltage vector u^* as a control input is not viable, as explained in Section 9.1. Dynamically superior is a direct modification of the actual pulse pattern.

An algorithm for this task is already implemented: the trajectory controller, described in Section 7.4. It permits the fastest possible manipulation of the stator flux linkage vector. Accordingly, the superimposed controllers must generate a signal Ψ_s^* as the controlling input to the modulator.

At rotor field orientation, Section 8.2.2, the imaginary component of the rotor flux $\Psi_{rq} = 0$. It is then only the real component of the stator flux linkage vector that controls the rotor flux magnitude Ψ_{rd}. This is established by the rotor equation (2.9b), page 12, as

$$\tau_r' \frac{d\Psi_{rd}}{d\tau} + \Psi_{rd} = k_s \Psi_{sd} . \qquad (9.4)$$

The reference signal Ψ_{sd}^* is obtained from the output of a rotor flux controller.

The electromagnetic torque is given by

$$T_e = \frac{k_r}{\sigma l_s} \Psi_{rd} \Psi_{sq} \qquad (9.5)$$

which defines Ψ_{sq}^* as the output of a speed controller.

Fig. 9.3 shows the control structure. The output signals of the rotor flux controller and of the velocity controller are joined to form the complex reference signal Ψ_s^*. This signal is compared with the actual fundamental flux vector $\hat{\Psi}_{s1}$ obtained from (9.2). The result is the error of the fundamental stator flux vector,

$$\Delta\Psi_{s1} = \Psi_s^* - \hat{\Psi}_{s1}, \qquad (9.6)$$

referred to in field coordinates. It represents the volt-second difference that occurs in a transient condition, [11]. The signal is transformed to stationary coordinates to become the error vector $\Delta\Psi_{s1}^{(S)}$.

The error vector is the input to the synchronous optimal pulsewidth modulator. It is added there to the dynamic modulation error $\hat{d}$ as shown in Fig. 9.4. The vector sum $\hat{d} + \Delta\Psi_{s1}$ is fed to the trajectory controller, which performs thus two tasks: (i) it eliminates the dynamic modula-

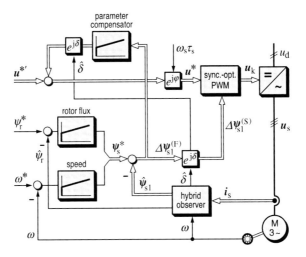

Fig. 9.3 The superimposed control system. The synchronous optimal pulse-width modulator is detailed in Fig. 9.4, and the hybrid observer in Fig. 9.2. The transformations between the stator and field coordinate systems are performed by the field angle $\hat{\delta}$.

tion error $\hat{d}$ defined by (7.14), page 83, and (ii) it executes the required dynamic changes of the drive control as commanded by the error vector $\Delta\Psi_{s1}$. The pattern modification process is described in detail in Section 7.4.1. The error vector $\hat{d} + \Delta\Psi_{s1}$ is minimized by, either advancing in time, or delaying, the switching transitions of the prevailing pulse pattern $P(m, N)$. Torque and flux control is thus achieved by a direct manipulation of the pulse pattern. The commanded changes are superimposed on the optimal pattern. As the optimum target trajectory Ψ_{ss} is referred to in this process, Fig. 9.4, the effect is a pattern optimization in real-time, [11].

The manipulation of the pulse pattern is a nonlinear process. The approach fully exploits the voltage margin of the inverter. The resulting response is therefore of the deadbeat type. Such performance cannot be obtained by linear controllers.

123

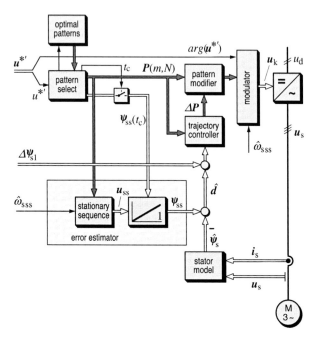

Fig. 9.4 Signal flow graph of the stator flux trajectory controller. The vector sum $\hat{d} + \Delta\psi_{s1}$ is fed to the trajectory controller. A pattern modification ΔP is performed to eliminate the error in both the harmonic and in the fundamental trajectory of the stator flux linkage vector.

9.5 Compensation of estimation errors

As shown in Fig. 9.1, the first estimate $u^{*'}$ of the reference voltage vector u^* is generated using signals from the hybrid observer, Fig. 9.2. The estimate depends on various machine parameters that form part of the respective model equations. The parameters vary with machine temperature, load, and magnetization level.

The controlling signal $\hat{u}_1 = u^{*'}$ of the self-controlled machine may be therefore in error. Its compensation through the superimposed controllers for rotor flux and speed is undesirable as it would unnecessarily activate the trajectory controller.

A better approach consists in using a nonzero long-term average of

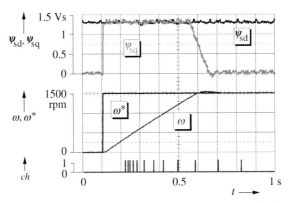

Fig. 9.5   Acceleration from standstill to rated speed; the marker signals ch indicate the instants where the pulse pattern changes.

the error vector $\Delta\Psi_{s1}$ to generate a correcting signal. It is obtained through a PI controller from that error, then transformed to stationary coordinates and added to u^{*}. As the PI controller is implemented in field coordinates, Fig. 9.3, it eliminates the offset components in $\Delta\Psi_{s1}$. These are predominantly caused by model parameter errors. Their slow variations in time justify to compensate only the long-term average of the error.

The integral time constant τ_i of the PI controller has a high value and the proportional gain g_c is small. The correction signal is thus limited in bandwidth so as to maintain the smoothness of the modulator input. It is advanced in phase angle by the transformation $e^{j\omega_s\tau_s}$, where τ_s is the normalized sampling interval of the control algorithm. The transformation is shown in Fig. 9.3; it compensates for the time delay of the discrete signal processing.

9.6   Performance evaluation

The low-voltage ac drive of the experimental setup is tested in various transient conditions, in order to evaluate the performance of the proposed closed-loop control scheme. Fig. 9.5 shows the response to a commanded step change of speed. The machine accelerates from standstill to rated speed within 0.5 s. Although the switching frequency is less than 200 Hz, an immediate response of the torque component Ψ_{sq} oc-

125

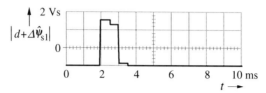

Fig. 9.6 Magnitude of the error vector $d + \Delta \Psi_{s1}$ following the step change in Fig. 9.5 (time scale 100 times enlarged). The error is eliminated within two sampling intervals.

curs. The field component Ψ_{sd} is maintained at nominal value without showing cross-coupling effects. The modulator operates permanently in transient conditions which entails numerous pulse pattern changes. Each pattern change is indicated by a marker signal ch in the bottom trace.

Fig. 9.6 illustrates the compensation of the dynamic error signal $d + \Delta \Psi_{s1}$ that results from the transient process Fig. 9.5. The time displacements of the switching transitions that compensate the error are calculated according to (7.23), page 90. The volt-second change takes immediate effect in the next sampling interval resulting in a deadbeat elimination of the transient. The effect on the respective stator current components i_d and i_q is illustrated in Fig. 9.7. The estimated fundamental current components $\hat{i}_{d1}$ and $\hat{i}_{q1}$ are also shown in this figure.

Fig. 9.8 shows the stator current trajectory with a step change applied from no-load to nominal load. It demonstrates the deadbeat behavior of the trajectory tracking. The algorithm is superior to conventional linear controllers in that it directly manipulates, and also optimizes, the pulse sequence in real-time.

Fig. 9.9 shows the estimated fundamental current components $\hat{i}_{d1}$ and $\hat{i}_{q1}$ of the same process immediately after the load change. The reference current components i_d* and i_q* are not the controlling variables. They are only derived from $i_s* = (1/l_\sigma)\Psi_s*$ for monitoring.

The deadbeat response is also confirmed by the stator flux trajectories in Fig. 9.10. Rise time and overshoot of the controlled variables are kept at a minimum during performing the large-signal step response from no-load to full load. Cross-coupling is almost zero. The trajectory controller makes the stator flux linkage vector independent of the machine dynamics. The reaction from the rotor to the stator is interrupted with the stator flux linkage vector independently forced.

126

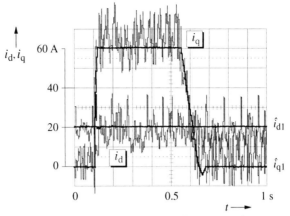

Fig. 9.7 Stator current components i_d and i_q of the acceleration process Fig. 9.5. The estimated current components $\hat{i}_{d1}$ and $\hat{i}_{q1}$ are also shown.

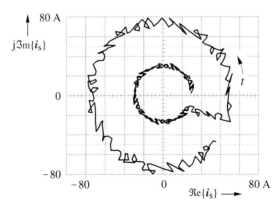

Fig. 9.8 Trajectory of the measured stator current vector i_s recorded at a step transition from no-load operation to nominal load.

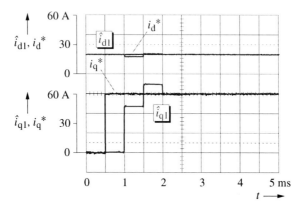

Fig. 9.9   Components of the reference current $i_s^* = i_d^* + ji_q^*$ and the fundamental stator current $\hat{i}_1 = \hat{i}_{d1} + j\hat{i}_{q1}$ during the transient process Fig. 9.8.

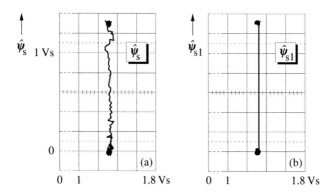

Fig. 9.10   Trajectories of the stator flux linkage vector during the transient process Fig. 9.8, (a) stator flux linkage, (b) estimated fundamental component.

128

10. Summary

The three-level inverter topology offers reduced harmonic distortion of the output current and low voltage stress of the semiconductor switches [1]. It enables operation at very low switching frequency and hence is a preferred solution for high-power, medium-voltage applications.

 Of particular industrial interest is the neutral point clamped (NPC) topology [2]. An intrinsic natural balancing mechanism of the neutral point clamped inverter topology eliminates long-term neutral point potential offsets. Transient conditions, however, may create successive increments of the offset to high values, which requires fast compensation.

The problem is especially pronounced when the inverter operates at very low switching frequency. The on-time durations of the switching states then increase, which entails higher values of the neutral point potential error.

A technique for fast elimination of the neutral point potential error in the low modulation range exploits the existence of two redundant subbridges in three-level inverters of the neutral point clamped topology. At steady-state, the two subbridges are activated in an alternating sequence, each executing a full modulation subcycle; the average neutral point current is then zero. A neutral point potential error generated at transient operation is eliminated by activating that particular subbridge that reduces the error. It is activated for as many subcycles as needed. All transitions are effected without additional commutations.

Hence the control entails no penalty. It reacts fast as it exploits always the maximum error gradient.

Setting the switching frequency to very low values allows minimizing the switching losses of the power semiconductors, but it has an immediate effect on the dynamic behavior of the system. It increases the signal delay in the control structure and intensifies the undesired cross-coupling between torque and flux in systems controlled in closed-loop. Standard feedforward compensation reduces the effect, but may not be considered satisfactory if the switching frequency is below 1 kHz.

A strategy to solve this problem relies on modelling the machine and the inverter without commonly employed approximations. This permits the design of a current controller with complex gain coefficients.

This controller permits achieving high dynamic performance, even when the switching frequency is restricted to very low values. However, the choice of the pulsewidth modulation method needs to be reevaluated: conventional space vector modulation contributes to high harmonic currents in the stator windings. The effect is particularly pronounced at high values of the modulation index.

Employing synchronous optimal modulation is a better solution. Assuming steady-state conditions, the method optimizes the pulse patterns so that the total distortion factor of the stator currents is minimum. Depending on the actual operating point, a particular pattern is retrieved from a memory of the controlling microprocessor and used for inverter control.

The improvement is significant in the modulation index range $m > 0.3$. The technique enables medium-voltage drives to operate at very low switching frequency down to 200 Hz. The switching losses of the power semiconductors are then reduced which permits increasing the maximum load current of the inverter.

Off-line optimization of the PWM sequence is only valid at steady-state. Even small changes of the modulation index, as in quasi steady-state operation, invoke transitions between neighboring pulse patterns. The optimization of the switching sequences is lost: undesired harmonic excursions appear as a result. The dynamic modulation error $\hat{d}$ serves as a means to measure the magnitude and orientation in space of such excursions. The error vector represents the deviation of the stator flux linkage space vector from a target trajectory that is generated from the actual optimal steady-state pattern. Relating the definition of the dynamic modulation error on the stator flux linkage vector introduces insensitivity against variations of the machine parameters.

The harmonic excursions are avoided, or minimized in cases, by forcing the estimated space vector of the stator flux linkage to follow the optimum, steady-state flux trajectory. To achieve this, the switching transitions in the pulse pattern in actual use are displaced such that the magnitude of the dynamic modulation error reduces to zero.

Stator flux trajectory tracking accounts for an optimization of the pulse patterns in real-time; it permits operation of the machine drive at quasi steady-state. To achieve high dynamic performance, control of the machine torque is imposed in closed-loop. It requires the estimation of the instantaneous value of the fundamental stator current as a feedback signal.

Two types of observers were analyzed for this purpose. The classical full-order observer introduces a dynamic phase lag of two sampling intervals when excited by a Dirac impulse test signal. The estimated current signal at closed loop control contains additional distortions. Better results are achieved using a hybrid observer that models the stator winding in stationary coordinates and the rotor winding in field coordinates. This observer reproduces the fundamental current without phase displacement or distortion.

Also the space vector of the fundamental stator flux is estimated at the output of the hybrid observer. It serves to create a replica of the fundamental component $\hat{u}_1$ of the stator voltages at the machine terminals. This signal is used as the controlling input to the pulsewidth modulator, $u^* = \hat{u}_1$. The method allows the system to perpetuate any given operating point by reestablishing the machine voltages through the modulator and the inverter. The machine is thus self-controlled. It allows the pulsewidth modulator to meet the requirement of optimal steady-state operation.

The self-controlled machine has no external control input. To achieve controllability, a control input is provided to the system by forming the error $\Delta\Psi_{s1}$ between a reference flux vector $\Psi_s{}^*$ and the estimated fundamental $\hat{\Psi}_{s1}$. The error $\Delta\Psi_{s1}$ is added to the dynamic modulation error $\hat{d}$ and fed to the trajectory controller. Thus, an existing torque command influences directly the switching transitions of the pulse patterns. While the trajectory controller achieves deadbeat response and eliminates cross-coupling effects, the system is operated from the undistorted reference voltage; the steady-state condition of synchronous optimal modulation is thus maintained.

The method was developed on a 36-kVA prototype and subsequently implemented in an industrial 2.5 MVA drive system. Experimental results are provided to evaluate the performance of the aforementioned schemes.

References

1. J. Holtz, "Self-Commutated Three-Phase Inverters with Staircase Voltage Waveforms for High-Power Applications at Low Switching Frequency" (in German), *Siemens Forschungs- und Entwicklungsberichte*, Vol. 6, No. 3, Springer 1977, pp. 164-171.

2. A. Nabae, I. Takahashi and H. Akagi, "A New Neutral Point Clamped PWM Inverter", *IEEE Transactions on Industry Applications*, Vol. 17, No. 5, Sep./Oct. 1981, pp. 518-523.

3. H. du T. Mouton, "Natural Balancing of Three-Level Neutral-Point-Clamped PWM Inverters", *IEEE Transactions on Industrial Electronics*, Vol. 49, No. 5, Oct. 2002, pp. 1017-1025.

4. J. Holtz, J. Quan, J. Pontt, J. Rodriguez, P. Newman and M. Miranda, "Design of Fast and Robust Current Regulators for Medium Voltage Drives Based on Complex State Variables", *IEEE Transactions on Industry Applications*, Vol. 40, No. 5, Sep./Oct. 2004, pp. 1388-1397.

5. J. Holtz and N. Oikonomou, "Synchronous Optimal Pulsewidth Modulation and Stator Flux Trajectory Control for Medium Voltage Drives", *IEEE Transactions on Industry Applications*, Vol. 43, No. 2, Mar./Apr. 2007, pp. 600-608.

6. G. S. Buja, "Optimum Output Waveforms in PWM Inverters", *IEEE Transactions on Industry Applications*, Vol. 16, No. 6, Nov./Dec. 1980, pp. 830 -836.

7. J. Holtz and B. Beyer, "Fast Current Trajectory Control Based on Synchronous Optimal Pulsewidth Modulation", *IEEE Transactions on Industry Applications*, Vol. 31, No. 5, Sep./Oct. 1995, pp. 1110-1120.

8. J. Holtz and B. Beyer, "The Trajectory Tracking Approach – A New Method for Minimum Distortion PWM in Dynamic High-Power Drives", *IEEE Transactions on Industry Applications*, Vol. 30, No. 4, July/Aug. 1994, pp. 1048-1057.

9. B. Beyer, "Schnelle Stromregelung für Hochleistungsantriebe mit Vorgabe der Stromtrajektorie durch off-line optimierte Pulsmuster", Ph.-D. Thesis, Wuppertal University, 1998. ISBN 3-89653-462-9, Ed. Mainz Verlag, Aachen, Germany, 1999.

10. J. Holtz, "Pulsewidth Modulation for Electronic Power Converters", *Proceedings of the IEEE*, Vol. 82, No. 8, Aug. 1994, pp. 1194-1214.

11. N. Oikonomou and J. Holtz, "Stator Flux Trajectory Tracking Control for High-Performance Drives", *IEEE Industry Applications Society Annual Meeting*, Tampa, FL, Oct. 2006, pp. 1268-1275.

12. J. Holtz, "The Representation of AC Machine Dynamics by Complex Signal Flow Graphs", *IEEE Transactions on Industrial Electronics*, Vol. 42, No. 3, 1995, pp. 263-271.

13. ABB Power Semiconductor Devices Product Catalogue, "www.abb.ch/ProductGuide", *ABB Power Technologies*, Lenzburg, Switzerland, July 2008.

14. S. Bernet, "Recent Developments of High Power Converters for Industry and Traction Applications", *IEEE Transactions on Power Electronicss*, Vol. 15, No. 6, Nov. 2000, pp. 1102-1117.

15. E. Cengelci, S. U. Sulistijo, B. O. Woom, P. Enjeti, R. Teodorescu, and F. Blaabjerg, "A New Medium Voltage PWM Inverter Topology for Adjustable Speed Drives", *IEEE Industry Applications Society Annual Meeting*, St-Louis, MO, Oct. 1998, pp. 1416-1423.

16. T. A. Meynard and H. Foch, "Multi-level Choppers for High Voltage Applications", *European Power Electronic Drives Journal*, Vol. 2, No. 1, Mar. 1992, pp. 41.

17. C. Hochgraf, R. Lasseter, D. Divan, and T. A. Lipo, "Comparison of Multilevel Inverters for Static Var Compensation", *EEE Industry Applications Society Annual Meeting*, Denver, CO, Oct. 1994, pp. 921-928.

18. P. Hammond, "A New Approach to Enhance Power Quality for Medium Voltage Ac drives", *IEEE Transactions on Industry Applications*, Vol. 33, No. 1, Jan./Feb. 1997, pp. 202-208.

19. W. A. Hill and C. H. Harbourt, "Performance of Medium Voltage Multilevel Inverters", *IEEE Industry Applications Society Annual Meeting*, Phoenix, AZ, Oct. 1999, pp. 1186-1192.

20. S. Bernett, D. Krug, S. S. Fazel, nd K. Jalili, "Design and Comparison of 4.16 kV Neutral Point Clamped, Flying Capacitor and Series Connected H-Bridge Multi-Level Converters", *IEEE Industry Applications Society Annual Meeting*, Hong-Kong, Oct. 2005, pp. 121-128.

21. T. Brückner and D. G. Holmes, "Optimal Pulse-Width Modulation for Three-Level Inverters", *IEEE Transactions on Power Electronics*, Vol. 20, No. 1, Jan. 2005, pp. 82-89.

22. D. G. Holmes and T. A. Lipo, "Pulse Width Modulation for Power Converters: Principle and Practice", Ed. Wiley & Sons, ISBN-13: 978-9471208143, 2003.

23. A. Busse and J. Holtz, "Multiloop Control of a Unity Power Factor Fast-Switching AC to DC Converter", *IEEE Power Electronics Specialists Conference*, 1982 Record, Cambridge/Ma., pp. 171-179.

24. J. Pou, R. Pindado and D. Boroyevich, "Evaluation of the Low-Frequency Neutral-Point Voltage Oscillations in the Three-Level Inverter", *IEEE Transactions on Industrial Electronics*, Vol. 52, No. 6, Dec. 2005, pp. 1582-1588.

25. J. K. Steinke, "Switching Frequency Optimal PWM Control of a Three-Level Inverter", *IEEE Transactions on Power Electronics*, Vol. 7, No. 3, Jan. 1992, pp. 487-496.

26. F. Wang, "Multilevel PWM VSIs – Coordinated Control of Regenerative Three-Level Neutral Point Clamped Pulsewidth-Modulated Voltage Source Inverters", *IEEE Industry Applications Magazine*, July/Aug. 2004, pp. 51-58.

27. J.-O. Krah and J. Holtz, "High-Performance Current Regulation and Efficient PWM Implementation for Low Inductance Servo Motors", *IEEE Transactions on Industry Applications*, Vol. 35, No. 5, Sep./Oct. 1999, pp. 1039-1049.

28. Pablo-F. Lara-Reyes, "Trajectory Tracking of Precalculated Patterns for Current Source Rectifiers", Ph.-D. Thesis, Wuppertal University, 2006, ISBN 3-86130-777-4, Ed. Mainz Verlag, Aachen, Germany, 2006.

29. A. M. Khambadkone, "A Study of Fast Digital Current Control Methods for High Power Industrial Drives using Field Orientation", Ph.-D. Thesis, Wuppertal University, 1995. Ed. Mainz Verlag, Aachen, Germany, 1995.

30. S. Ogasawara, H. Akagi and A. Nabae, "A Novel PWM Scheme of Voltage Source Inverters based on Space Vectors", *EPE European Conference on Power Electronics and Applications*, Aachen 1989, pp. 2297-1202.

31. H. Kleinrath, "Stromrichtergespeiste Drehfeldmaschinen", Ed. Springer-Verlag KG, ISBN 3-21181-565-1.

32. B. N. Psenicnyi and J. M. Danilin, "Numerische Methoden für Extremalaufgaben", *VEB Deutscher Verlag der Wissenschaften*, Berlin, 1982.

33. J. Holtz, W. Lotzkat, and A. Khambadkone, "On Continuous Control of PWM Inverters in the Overmodulation Range with Transition to the Six-Step Mode", *IEEE Transactions on Power Electronics*, Vol. 8, No. 4, Oct. 1993, pp. 546-553.

34. F. Zach, R. Martinez, S. Keplinger, and A. Seiser, "Dynamical Optimal Switching Patterns for PWM Inverter Drives (for Minimization of the Torque and Speed Ripples)", *IEEE Transactions on Industry Applications*, Vol. IA, No. 4, 1985, pp. 975-986.

35. A. K. Gupta and A. M. Khambadkone, "A General Space Vector PWM Algorithm for Multilevel Inverters, Including Operation in the Overmodulation Range", *IEEE Transactions on Power Electronics*, Vol. 22, No. 2, Mar. 2007, pp. 517-526.

36. Eupec Device Simulator, http://www.eupec.com/p/iposim/iposim.htm

37. J. Holtz, "Sensorless Control of Induction Motor Drives", *Proceedings of the IEEE*, Vol. 90, No. 8, Aug. 2002, pp. 1359-1394.

38. M. Hinkkannen and J. Luomi, "Modified Integrator for Voltage Model Flux Estimation of Induction Motors", *IEEE Transactions on Industrial Electronics*, Vol. 50, No. 4, Aug. 2003, pp. 818-820.

39. J. Holtz and J. Quan, "Drift and Parameter Compensated Flux Estimator for Persisten Zero Stator Frequency Operation of Sensorless Controlled Induction Motors", *IEEE Transactions on Industry Applications*, Vol. 39, No. 4, July/Aug. 2003, pp. 1052-1060.

40. M. Hinkkannen, "Flux Estimators for Speed-Sensorless Induction Motor Drives", Ph.-D. Thesis, Helsinki University of Technology, 2004. ISBN 951-22-7188-5, Espoo, Finland, 2004.

41. J. Holtz and J. Quan, "Sensorless Vector Control of Induction Motors at Very Low Speed using a Nonlinear Inverter Model and Parameter Identification", *IEEE Transactions on Industry Applications*, Vol. 38, No. 4, July/Aug. 2002, pp. 1087-1095.

42. D. G. Luenberger, "An introduction to observers", *IEEE Transactions on Automatic Control*, Vol. AC-16, No. 6, Dec. 1971, pp. 596-602.

43. H. Kubota, K. Matsuse, and T. Nakano, "DSP Based Speed Adaptive Flux Observer of Induction Motor", *IEEE Transactions on Industry Applications*, Vol. 29, No. 2, Mar./Apr. 1993, pp. 344-348.

44. J. Holtz, "On the Spatial Propagation of Transient Magnetic Fields in AC Machines", *IEEE Transactions on Industry Applications*, Vol. 32, No. 4, July/Aug. 1996, pp. 927-937.

45. H. Kubota, K. Matsuse, and T. Nakano, "Speed Sensorless Field-Oriented Control of Induction Motor with Rotor Resistance Adaptation", *IEEE Transactions on Industry Applications*, Vol. 30, No. 5, Sept./Oct. 1994, pp. 1219-1224.

46. G. S. Buja and G. B. Indri, "Optimal Pulsewidth Modulation for Feeding AC Motors", *IEEE Transactions on Industry Applications,* Vol. 13, No. 1, Jan./Feb. 1977, pp. 38-44.

Appendix

Normalization

The nominal per-phase stator voltage at star connection $U_{phR} = U_R/\sqrt{3}$, is chosen as base value for the normalization of voltage variables. In the previous equation, U_R is the rated phase-to-phase voltage. Similarly, the nominal per-phase stator current at star connection $I_{phR} = I_R$ is the base value for the normalization of current variables.

The following table lists the normalization values for all variables and machine parameters used throughout the work.

voltage	$U_R/\sqrt{3}$	mechanical speed	ω_{sR}/p
current	I_R	torque	$\sqrt{3}pU_RI_R/\omega_{sR}$
power	$\sqrt{3}U_RI_R$	resistance	$U_R/(\sqrt{3}I_R)$
flux linkage	$U_R/(\sqrt{3}\omega_{sR})$	inductance	$U_R/(\sqrt{3}\omega_{sR}I_R)$

Time is normalized as $\tau = \omega_{sR}t$, where ω_{sR} is the rated stator frequency.

Machine data

Table 1 - medium-voltage induction machine

rated power	P_R	= 1800 kW	
rated voltage	U_R	= 4.16 kV	
rated current	I_R	= 300 A	
rated power factor	$cos\phi$	= 0.84	
rated stator frequency	ω_{sR}	= $2\pi\cdot60$ r/s	
rated angular velocity	n_R	= 1790 rpm	
stator resistance	r_s	= 69.7 mΩ	0.009 pu
rotor resistance	r_r	= 35.3 mΩ	0.004 pu
stator inductance	l_s	= 89.4 mH	4.203 pu
rotor inductance	l_r	= 90.0 mH	4.231 pu
stator leakage inductance	$l_{s\sigma}$	= 1.9 mH	0.089 pu
rotor leakage inductance	$l_{r\sigma}$	= 2.5 mH	0.118 pu

Table 2 - low-voltage induction machine

rated power	P_R	= 30 kW
rated voltage	U_R	= 380 V
rated current	I_R	= 60 A
rated power factor	$cos\phi$	= 0.83

rated stator frequency	ω_{sR}	= $2\pi \cdot 50$ r/s	
rated angular velocity	n_R	= 1465 rpm	
stator resistance	r_s	= 150 mΩ	0.041 pu
rotor resistance	r_r	= 143 mΩ	0.039 pu
stator inductance	l_s	= 20.27 mH	1.739 pu
rotor inductance	l_r	= 20.49 mH	1.757 pu
stator leakage inductance	$l_{s\sigma}$	= 0.77 mH	0.066 pu
rotor leakage inductance	$l_{r\sigma}$	= 0.99 mH	0.085 pu

List of symbols

Space vectors

A	current density distribution
d	dynamic modulation error
i	current
u	voltage
ψ	flux linkage

Scalar variables and parameters

d	i) magnitude of dynamic modulation error vector
	ii) distortion factor
f	frequency (Hz)
g	proportional gain
I	current magnitude (A)
i	current magnitude (pu)
k	coupling factor (pu)
l	inductance (pu)
m	modulation index
N	pulse number
n	rotor angular velociry (rpm)
p	pole pair number
r	resistance (pu)
s	phase state
T	torque (pu)
t	time (s)
U	voltage magnitude (V)
u	voltage magnitude (pu)
δ	field angle (rad)
ε	design parameter
σ	leakage factor (pu)
τ	time (pu)

ψ	flux linkage magnitude (pu)
ω	frequency (r/s)

Variables in the frequency domain

D	denominator of transfer function
F	transfer function
N	nominator of transfer function
s	Laplace operator

Complex variables

G	correction tensor
I	fundamental content of periodic current
λ	eigenvalue

Subscripts

X_0	subcycle
x_1	fundamental
x_a, x_b, x_c	phase a, b, c
X_c	closed-loop
X_d	delay
x_d (x_q)	real (imaginary)
X_e	electromagnetic
X_{FL}	full load
x_h	harmonic
x_i	i) current
	ii) induced
x_k	switching state vector
X_L	load
X_m	machine
x_m	mechanical
x_{max}	maximum
x_{min}	minimum
x_n	neutral point
X_{NL}	no-load
X_o	open-loop
X_p	plant
x_{per}	periodic
x_{ph}	phase
X_r	controller
x_r	rotor
x_R	rated

X_{S}	sampling
x_{S}	stator
x_{S}	switching
x_{ss}	steady-state
x_{rms}	rms
x_{u}	voltage
x_{Y}	common connection terminal
x_{σ}	leakage

Superscripts

$x^{(\mathrm{F})}$	field coordinate system
$x^{(\mathrm{S})}$	stationary coordinate system
$\tilde{X}$	Laplace transform
$\tilde{x}$	predicted value
x'	i) transient value
	ii) first estimate
x^*	reference value
$\check{x}$	transient deviation
$\bar{x}$	time-average
$\hat{x}$	estimated value
$\hat{x}$	peak value

Operations

$arg(x)$	phase angle of vector x
$dx/d\tau$	time derivative of variable x
$exp(x)$	exponential of variable x
Δx	difference of variable x
$\Re\{x\}$	real part of vector x
$\Im\{x\}$	imaginary part of vector x
$x \times y$	external product of vector x with vector y
$x \circ y$	internal product of vector x with vector y
$x\big|_{z}$	projection of vector x on the z-axis
$x \in S$	element x belongs to set S

Abbreviations

pu	per unit
NPC	neutral point clamped
PWM	pulsewidth modulation
SOM	synchronous optimal modulation
SVM	space vector modulation